Bridging Eigenvalue Theory and Practice - Applications in Modern Engineering

Edited by Bruno Carpentieri

Published in London, United Kingdom

Bridging Eigenvalue Theory and Practice - Applications in Modern Engineering
http://dx.doi.org/10.5772/intechopen.1003394
Edited by Bruno Carpentieri

Contributors
Abhijith Ajayakumar, Azwirman Gusrialdi, Bruno Carpentieri, Carlo Grillenzoni, Dawit Hiluf Hailu, Deependra Neupane, Kenneth McDonald, Marcela Parraguez, Marcus Carlsson, Mudassir Shams, Nawaraj Poudel, Raju K. George, Raúl Jiménez, Tor A. Kwembe, Zhihua Qu

First published in London, United Kingdom, 2025 by IntechOpen
IntechOpen is the global imprint of INTECHOPEN LIMITED, registered in England and Wales, registration number: 11086078, 167-169 Great Portland Street, London, W1W 5PF, United Kingdom

For EU product safety concerns: IN TECH d.o.o., Prolaz Marije Krucifikse Kozulić 3, 51000 Rijeka, Croatia, info@intechopen.com or visit our website at intechopen.com.

British Library Cataloguing-in-Publication Data
A catalogue record for this book is available from the British Library

Bridging Eigenvalue Theory and Practice - Applications in Modern Engineering
Edited by Bruno Carpentieri
p. cm.
Print ISBN 978-1-83634-248-9
Online ISBN 978-1-83634-247-2
eBook (PDF) ISBN 978-1-83634-249-6

Meet the editor

Bruno Carpentieri received his Laurea in Applied Mathematics from Bari University, Italy, and a Ph.D. in Computer Science from INPT, France. He held postdoctoral positions in Austria and Italy before joining academia. He has been an Assistant Professor at the University of Groningen in the Netherlands, a Reader at Nottingham Trent University in England, and, since 2017, an Associate Professor in Applied Mathematics at the Free University of Bozen-Bolzano in Italy. His research focuses on numerical linear algebra, high-performance computing, and computational mathematics, with applications in scientific modeling and engineering. He has served on scientific advisory boards, editorial committees, and review panels for leading journals and conferences. He has supervised over 20 student projects at various levels and authored about 80 peer-reviewed publications.

Contents

Preface

Eigenvalue theory is a fundamental pillar of applied mathematics, providing essential tools for analyzing the stability, controllability, and structural properties of complex systems. From classical linear algebra to modern engineering applications, eigenvalues and eigenvectors play a crucial role in different domains, including control theory, quantum mechanics, data science, and high-performance computing. As engineering and computational sciences continue to evolve, new challenges and applications emerge, demanding innovative numerical methods and computational strategies.

This volume, *Bridging Eigenvalue Theory and Practice – Applications in Modern Engineering*, brings together a collection of contributions that explore both the theoretical foundations and practical implementations of eigenvalue analysis. The chapters present a balanced perspective, ranging from fundamental mathematical insights to real-world engineering and computational science applications.

The book begins with Chapter 1, which offers a pedagogical perspective on eigenvalues in two-dimensional spaces, emphasizing geometric interpretations and invariant subspaces to enhance conceptual understanding. Chapter 2 builds on this foundation, exploring the role of eigenvalues and eigenvectors in controllability analysis, with applications in linear and nonlinear control systems. Chapter 3 then introduces matrix exponentiation techniques using eigenvalues, focusing on applications in quantum mechanics and dynamical systems.

A deeper mathematical treatment follows in Chapter 4, which delves into the spectral perturbation theory of Hermitian matrices, addressing fundamental challenges in dealing with degenerate eigenvalues. Chapter 5 presents a hybrid parallel scheme for solving nonlinear eigenvalue problems using multiplicative calculus, improving computational efficiency and accuracy for engineering applications.

The transition to large-scale and data-driven approaches is marked by Chapter 6, which explores the eigen-analysis of multi-agent and large-scale systems using machine learning algorithms. This theme continues in Chapter 7, where eigenvalue methods are applied to mitigate multicollinearity in machine learning and high-dimensional data analysis, with a detailed discussion on Principal Component Analysis (PCA).

The final chapters highlight engineering applications of eigenvalue-based techniques. Chapter 8 presents the use of principal components and factor models for space-time data in remote sensing, addressing challenges in multidimensional image analysis. The book concludes with Chapter 9, which examines small-signal stability analysis of virtual impedance-based parallel inverters, demonstrating the impact of eigenvalue analysis in electrical power systems.

Through these contributions, this volume offers a comprehensive exploration of eigenvalue theory's role in contemporary science and engineering. By bridging fundamental theory with computational advancements and engineering applications, it serves as a valuable resource for researchers, practitioners, and students interested in the interplay between spectral methods and modern technology.

I would like to extend my sincere gratitude to all the authors for their valuable contributions, as well as to the reviewers whose insightful feedback has helped refine the content of this book. Special thanks go to the Publishing Process Manager, Mrs Ivana Barac, for her invaluable support and expertise throughout the editorial process.

We hope this volume inspires further research and innovation in eigenvalue-based methodologies and their applications across scientific and engineering disciplines.

Bruno Carpentieri
Faculty of Engineering,
Free University of Bozen-Bolzano,
Bolzano, Italy

Chapter 1

Perspective Chapter: The Concept of Eigenvalue in 2-Dimensional Spaces and Its Connections with Geometric Scaling and Invariant Subspace

Marcela Parraguez and Raúl Jiménez

Abstract

Eigenvalues and eigenvectors are typically introduced mechanically in introductory Linear Algebra courses in fields, such as engineering, science, and mathematics education. An alternative teaching method emphasizes a conceptual understanding by linking these concepts to one-dimensional invariant subspaces and geometric vector scaling. This pedagogical shift fosters a deeper comprehension by highlighting practical applications. By focusing on geometric interpretations, this approach not only aids university students in grasping abstract concepts but also demonstrates their practical relevance, effectively sidestepping the dense formalism usually associated with Linear Algebra. Encouraging students to visualize transformations and their effects holistically contributes significantly to their mathematical maturity and problem-solving skills.

Keywords: vector space, eigenvalues, geometric scaling, invariant subspace, linear transformation

1. Introduction

The concept of eigenvalue is undoubtedly one of the most widely applied principles in Linear Algebra. Its applications are well known in structural dynamics, graph theory, economics, the theory of equations, and stability, among others [1–4]. This concept is strongly linked to that of linear transformation and eigenvector because it defines an invariant direction of the eigenvector, meaning that the vector does not change direction when the linear transformation is applied to it. The role of the eigenvalue for this invariant direction is to "scale" the eigenvector without altering its direction and, in some cases, its orientation. In the scenario described above, the concepts of linear transformation, eigenvalue, and eigenvector are strongly related. We will now discuss the concept of eigenvalue and its connections according to **Figure 1**.

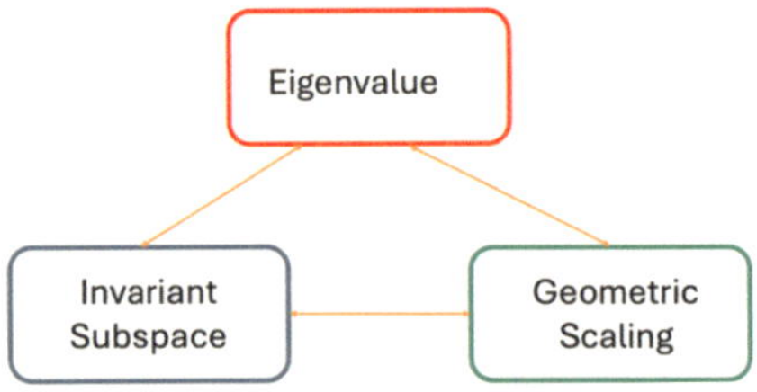

Figure 1.
Concepts interacting with eigenvalue.

Specifically, in this chapter we are interested to show the construction of eigenvalues in $\mathbb{R}^2$ connected with geometric scaling and one-dimensional invariant subspaces.

2. Theoretical framework and literature review

A useful group of concepts in Linear Algebra is eigentheory as mentioned in Carrel [5] and Wawro et al. [6], that is, the study of eigenvectors, eigenvalues,[1] and eigenspaces. We mention eigentheory because it is a set of conceptually complex notions, which are built from the understanding of multiple key ideas for Linear Algebra: linear transformation, vector spaces, basis of a vector space, matrix equation, determinant of a matrix, characteristic polynomial, and kernel of a matrix, among others. Investigating the construction of eigentheory in students—who learn—and teachers—who teach—these topics contribute to what is known about how teachers and learners conceptualize eigentheory [6–11].

This chapter addresses an aspect of eigentheory that has received relatively little attention in higher education research, namely, the interaction between eigenvalues, geometric scaling, and associated invariant subspaces.[2] Below is an example in $\mathbb{R}^2$, where the three previous concepts are explicit.

Consider the linear transformation $T : \mathbb{R}^2 \to \mathbb{R}^2$, which describes a symmetry respect to a fixed line L passing through the origin, as shown in **Figure 2**. This linear transformation has the line L as an invariant subspace. Besides, the line L' perpendicular to L and passing through the origin is also an invariant vector subspace. Therefore, this linear transformation has only two non-trivial invariant subspaces, which are L and L'. We observe that a matrix associated with T with respect to a certain base B is $(T)_B = \begin{pmatrix} 1 & 0 \\ 0 & -1 \end{pmatrix}$ whose eigenvalues are $\lambda_1 = -1$ and $\lambda_2 = 1$, which scale geometrically to the eigenvectors $v_1 = (1, 0)$ and $v_2 = (0, 1)$, respectively.

---

[1] Let $T: \mathbb{R}^2 \to \mathbb{R}^2$ be a linear transformation. A scalar $\lambda \in \mathbb{R}$ is called eigenvalue of T if and only if there exists $v, v \neq 0$, $v \in \mathbb{R}^2$, such that $T(v) = \lambda v$. If λ is an eigenvalue of T, the vector $v \in \mathbb{R}^2$ is called the eigenvector of T associated with the eigenvalue λ.

[2] Let V be a vector space, $S \leq V$ and $v \in S \Longrightarrow T(v) \in S$, where T is an endomorphism in V.

DOI: http://dx.doi.org/10.5772/intechopen.1007590

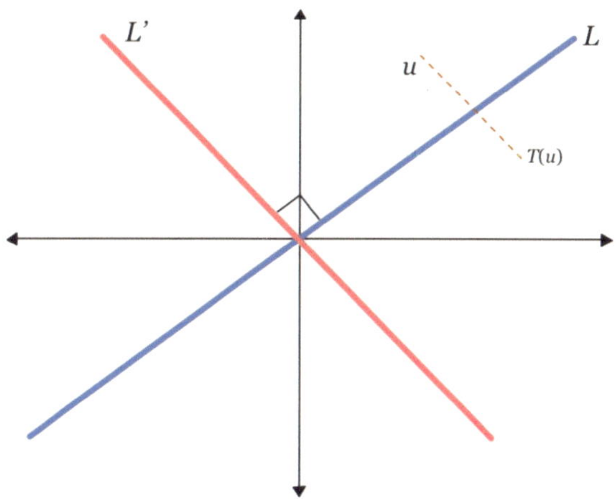

Figure 2.
Geometric representation of L *and* L' invariant subspaces of $\mathbb{R}^2$.

3. Methodology: Content analysis

The data for this study come from the work *Content Analysis methodology: An introduction to its methodology* [12], which we have used for: (1) the presentation of eigenvalue in an introductory one- or two semesters in Linear Algebra for undergraduates, as opposed to the proposal to present (2) eigenvalues via invariant subspaces of $\mathbb{R}^2$.

This comparison of two different ways of constructing eigenvalues will provide us with two paths of construction, the first through the classical concept and that involves scaling of the eigenvalue and another that involves the concept of invariant subspace.

The indicators to carry out this study and to be able to classify the mathematical objects that make possible the interaction in a systemic way between eigenvalues, geometric scaling, and associated invariant subspaces will be given via the matrix/symbolic, vectorial/symbolic, functional/symbolic, and geometric/symbolic algebraic representations, which are shown in **Table 1**, where A represents the matrix associated with the linear transformation T on a vector space V, λ eigenvalue of A, and S invariant vector subspace of V.

As a study technique, a content analysis, as mentioned in Krippendorff [12], was made, which considered the definition of eigenvalue, the invariant subspace, the

	Matrix/symbolic algebraic representations		
Scalar λ	$Ax = \lambda x$	$(A - \lambda I)x = 0$	$\det(A - \lambda I) = 0$
	Vector/symbolic representations		
λv	$Av = \lambda v$	$(A - \lambda I)v = 0$	$v \in S \rightarrow Av \in S$
	Functional/symbolic representations		
$T(v)$	$T(v) = \lambda v$	$(T - \lambda I)v = 0$	$v \in S \rightarrow T(v) \in S$
	Geometric/symbolic representations		
Dilatation	Parallelism of homothecy	Null projection	Invariant direction

Table 1.
Indicators via representations of the interaction between eigenvalues, geometric scaling, and invariant subspaces.

Research question	Path of inquiry	Categories
Problems regarding the interaction with eigenvalues and other concepts in Linear Algebra	1. The eigentheory as a theoretical framework How are the eigenvalues and other concepts of the interaction interpreted?	1.1 Theoretical proposal
		1.2 Justification
	2. Methodology framework How do we access the interaction between eigenvalues and other concepts?	2.1 Eigenvalue
		2.2 Invariant subspace
		2.3 Geometric scaling
		2.4 Analysis techniques
		2.5 Outcomes' presentation

Table 2.
Deductive categorization.

multiplication of a vector by a scalar, and analysis of Linear Algebra books included in the undergraduate bibliography courses on these topics, since, according to Mayring [13], a system of categories constitutes the central instrument of content analysis.

The categories were established in two stages in light of the objective of this book chapter. First, a deductive or top-down categorization was carried out [14], in which the definition of the *path of inquiry* was considered as that theoretical-methodological approach that allows us to address a study problem [15]. The outcomes are mentioned in **Table 2**.

Categories	Vector space $\mathbb{R}^2$
1.1 Theoretical proposal	1.1.1 Related concepts 1.1.2 Specific definitions of Linear Algebra 1.1.3 Mathematical representations 1.1.4 Examples that illustrate vector space
1.2 Justification	1.2.1 Specificity of the examples 1.2.2 Analysis of the types of representations
2.1 Eigenvalue	2.1.1 Definition 2.1.2 Representations
2.2 Invariant subspace	2.2.1 Definition 2.2.2 Representations
2.3 Geometric scaling	2.3.1 Definition 2.3.2 Representations
2.4 Analysis techniques	2.4.1 Focus on representations and arguments used 2.4.2 Colours diagram
2.5 Outcomes' presentation	2.5.1 New approach 2.5.2 Geometric interpretation 2.5.3 Projections 2.5.4 Engineering applications

Table 3.
Categories applied to the concept of eigenvalue in $\mathbb{R}^2$. Interaction indicators.

From the categories mentioned in **Table 2**, in a second-stage linear algebra textbooks were analyzed, and examples were constructed to show how the concept of eigenvalue, in the vector space $\mathbb{R}^2$, interacts with invariant subspace and geometric scaling concepts, which resulted in **Table 3**, in which indicators were obtained that were called *interaction indicators*.

Based on the interaction indicators given in **Table 3**, the following sections are presented.

4. Presentation of eigenvalue in an introductory Linear Algebra undergraduate course

The concept of eigenvalue is introduced in a classical undergraduate Linear Algebra course through a definition for a square matrix, preferably with coefficients in the field of the real numbers, as shown in **Figure 3**.

Once the definition of eigenvalue has been given, a theorem (**Figure 4**) is presented, which provides the procedural tools of how the eigenvalues of a matrix are determined.

In the previous scenario, an example is shown, which represents what is generally presented in the classrooms of classical Linear Algebra courses to operationalize the calculative dimension of the Theorem cited in **Figure 4**.

Example: Determine the eigenvalues of $A = \begin{pmatrix} 1 & 2 \\ 3 & 2 \end{pmatrix} \in M_2(\mathbb{R})$.

Solution: The expression $\lambda \in \mathbb{R}$ is obtained from the following equation, $|A - \lambda I| = 0$, that is: $\begin{vmatrix} 1-\lambda & 2 \\ 3 & 2-\lambda \end{vmatrix} = 0$, from where it is obtained, via calculation of the determinant that $(1-\lambda)(2-\lambda) - 6 = 0$, which implies that $\lambda = 4 \ \vee \ \lambda = -1$. From

<u>**Definition**</u> Let A be a matrix of $n \times n$ order. A λ scalar is called eigenvalue of A if exists a non-zero vector x such that $Ax = \lambda x$. The vector x is called the corresponding eigenvector to λ of A.

Figure 3.
Definition of eigenvalue [16].

Theorem 7.1.2. Let A be a matrix of $n \times n$ order and λ a real number, then the next statements are equivalent:

a) λ is an eigenvalue of A
b) The sistem equations $(\lambda I - A)x = 0$ has non-trivial solutions
c) In $\mathbb{R}^n$, there exist a non-zero vector x such that $Ax = \lambda x$
d) λ is a solution of the characteristic equation $(\lambda I - A)x = 0$

Figure 4.
Theorem that allows us to work on the calculative dimension of eigenvalue [17].

the above calculation and according to the theorem cited in **Figure 4**, we then concluded that the eigenvalues associated with the matrix A are 4 and -1.

Based on the last example, the way in which eigenvalues are presented generally emphasizes the calculative aspect of the detriment of the interpretive dimension, which is often unconnected with other mathematical concepts. However, the introduction of eigenvalues through invariant subspaces facilitates a systemic integration between the calculations and their geometric interpretations. We propose that this approach presented in this chapter allows for a deeper, more systemic understanding of the representations and concepts that are presented in **Table 1** and that are necessary for a deeper understanding of the concept of eigenvalue, as will be demonstrated in the next section for $\mathbb{R}^2$.

5. Presentation of eigenvalue via invariant subspace in $\mathbb{R}^2$ and examples

In this section, we will address the concept of eigenvalue in $\mathbb{R}^2$ because of scaling and the invariant direction of its associated eigenvector. The linear transformations defined in the plane have the characteristic that when acting on a vector, they may or may not alter its direction. We will focus our attention on those linear transformations T for which there are vectors whose images do not change direction. We observe that not all linear transformations have this property. A classic example is the endomorphism rotation $T_\theta : \mathbb{R}^2 \to \mathbb{R}^2$, where θ is the rotation angle, whose matrix representation with respect to the canonical basis is the matrix $[A]_{T_\theta} = \begin{bmatrix} cos\theta & -sen\theta \\ sen\theta & cos\theta \end{bmatrix}$. Particularly for $\theta = 90^\circ$, this transformation changes the direction of any vector on which it acts, therefore, there are no vectors that maintain their directions under the action of T. This implies that this linear transformation T_{90° has no eigenvectors in $\mathbb{R}^2$ since no vector is transformed into a multiple of itself and, therefore, cannot have real eigenvalues. In fact, this matrix has complex eigenvalues given by $\lambda = i$ and $\lambda = -i$, with eigenvectors over $\mathbb{C}^2$ given by $\begin{bmatrix} 1 \\ -i \end{bmatrix}$ and $\begin{bmatrix} 1 \\ i \end{bmatrix}$, respectively.

In this case, complex eigenvalues modify the magnitude and orientation of the eigenvector. In effect, the eigenvalues of the matrix $[A]_{T_\theta}$ are given by the numbers $\lambda = r(cos\theta \pm isen\theta) = re^{\pm i\theta}$. If we assume that an associated eigenvector can be written in the polar form $v_1 = r_1 e^{i\theta_1}$, on the equality $Av_1 = \lambda v_1$, the product $\lambda v_1 = re^{\pm i\theta} r_1 e^{i\theta_1} = rr_1 e^{i(\theta_1 \pm \theta)}$ shows us that the eigenvector v_1 is scaled (dilated or contracted) and undergoes a rotation. From now on, we will consider the case of real eigenvalues and eigenvectors, from which in equality $\lambda = r(cos\theta + isen\theta) = 0$ the only possibility that they are real is if it were true that $sen\theta = 0$. Therefore, eigenvectors are real for $\theta = k\pi, k \in \mathbb{Z}$, that is, for example, 180° rotations, which correspond to a change of direction but not of direction of the eigenvector.

We can then establish that a non-zero vector v is a vector proper of a linear transformation T if under the action of T, the vector does not change direction. From this, v is a proper vector of T, if v and its image $T(v)$ have the same direction. We can also say that $v \neq 0$ is a vector proper of $T \Leftrightarrow v$ is parallel with $T(v) \Leftrightarrow$ exists $\lambda \in \mathbb{R}$, such that $T(v) = \lambda\, v$.

This last equality suggests that we can pose the concept of eigenvector in terms of the concept of generated space. Indeed, the expression $T(v) = \lambda v$ is equivalent to say

$T(v) \in \langle v \rangle$ because the vectors of $\langle v \rangle$ subspace are multiples of v. From this statement, the role of the scalar λ is highlighted from two points of view: First, from the algebraic point of view, it introduces the concept of eigenvalue associated with the eigenvector v, and in turn, allows us to establish that the vector $T(v)$ belongs to the subspace generated by v. Second, from the geometric point of view, equality $T(v) = \lambda v$ establishes that λ fulfills a scaling function that can mean a dilation or contraction of the vector v without losing its direction. It makes sense then to label this address as an *invariant address* under T. Thus, we can establish that if λ is an eigenvalue associated with the eigenvector v, then:

$$w \in \langle v \rangle \Rightarrow w = kv \Rightarrow T(w) = T(kv) = kT(v) = k\lambda v = tv, \text{con } t = k\lambda. \tag{1}$$

Or $w \in \langle v \rangle \Rightarrow T(w) \in \langle v \rangle$ tells us that the vectors belonging to $W = \langle v \rangle$ remain in W under the action of T. This property of $\langle v \rangle$ as a subspace satisfies a more general definition, which is stated below.

Definition 1: Let V be a vector space. Let $T \in L(V) = \{T : V \to V/\text{a linear transformation}\}$. A subspace W is called invariant under T ($T-$ invariant) if $\forall r \in W$ we have $T(r) \in W$ [18].

From Definition 1, it can be deduced that $\langle v \rangle$ is a one-dimensional $T-$invariant space. We can thus state the following theorem.

Theorem 1: Let $T \in L(V)$. Let $v \neq 0$ and $W = \langle v \rangle$. T has an eigenvalue if and only if W is $T-$ invariant.

The justification for the statement of Theorem 1 follows from $W = \langle v \rangle$ is one-dimensional. Therefore, if W is invariant, then $T(v) \in W = \langle v \rangle$. This means that there exists λ, such that $T(v) = \lambda v$ from which T has an eigenvalue. Conversely, if λ is an eigenvalue of T, there exists $v \neq 0$, such that $T(v) = \lambda v$. Then, from (1), we have $W = \langle v \rangle$ is $T-$ invariant.

According to Definition 1 and Theorem 1, we can say that every line in the plane passing through the origin is an invariant one-dimensional subspace of a linear transformation in $\mathbb{R}^2$.

We recall below the definition of eigenspace associated with an eigenvalue, to enunciate an associated theorem that links the concept of eigenvalue with invariant subspace.

Definition 2: Let $T \in L(V)$. The eigenspace defined by an eigenvalue is given by $S_\lambda = \{v \in V : T(v) = \lambda v\}$.

Theorem 2: Let $T \in L(V)$. If λ is an eigenvalue of T then S_λ is an $T-$invariant subspace.

In the context of the above scenario, two examples are presented, to highlight the geometric scaling and the invariant subspace associated with its eigenvalue.

Example 1. Let $A = \begin{pmatrix} -3 & 5 \\ -2 & 4 \end{pmatrix} \in M_2(\mathbb{R})$ whose eigenvalues are $\lambda_1 = 2$ and $\lambda_2 = -1$. The eigenspaces are $S_{\lambda_1} = \langle v_1 \rangle = \langle (1,1) \rangle$ and $S_{\lambda_2} = \langle v_2 \rangle = \left\langle \left(\frac{5}{2}, 1\right) \right\rangle$. We can see from **Figure 5** that these subspaces are invariant under transformation $T(v) = Av$ and that they maintain their direction.

Example 2. Consider the following linear transformation $T : \mathbb{R}^2 \to \mathbb{R}^2$ represented with respect to the canonical basis by the matrix $A = \begin{pmatrix} -1 & 0 \\ 1 & 1 \end{pmatrix}$, whose eigenvalues are $\lambda_1 = 1$ and $\lambda_2 = -1$, with eigenspaces $S_{\lambda_1} = \langle (0,1) \rangle$ and $S_{\lambda_2} = \langle (-2,1) \rangle$. **Figure 6** shows these invariant subspaces.

Figure 6 shows the following.

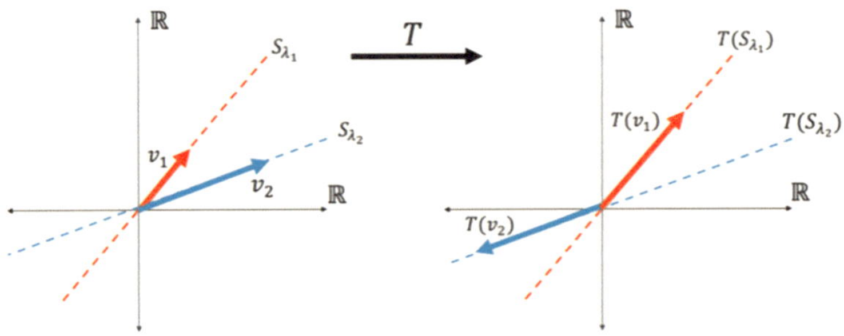

Figure 5.
$S_{\lambda_1} = \langle(1, 1)\rangle$ and $S_{\lambda_2} = \langle(\frac{5}{2}, 1)\rangle$ are invariant subspaces of $\mathbb{R}^2$.

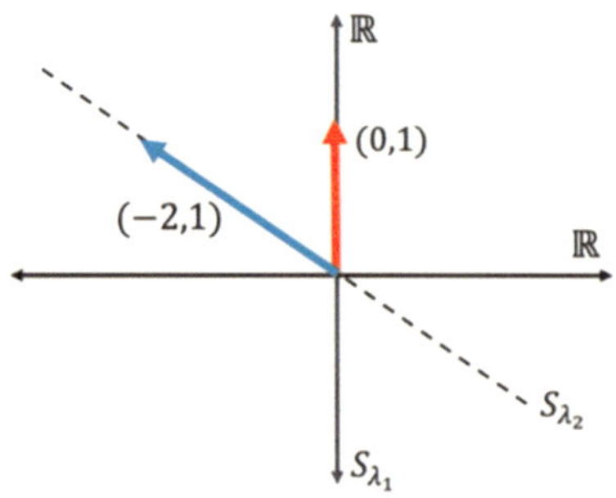

Figure 6.
$S_{\lambda_1} = \langle(0, 1)\rangle$ and $S_{\lambda_2} = \langle(-2, 1)\rangle$ are invariant subspaces of $\mathbb{R}^2$.

First: The vector $(-2, 1)$ is located in the second quadrant and belongs to the line defined by S_{λ_2}, so any multiple of this vector has the same direction and under the action of T, any of these vectors remains on the line S_{λ_2} since:

$$k(-2,1) \in S_{\lambda_2} \Rightarrow T(k(-2,1)) = kT(-2,1) = -k(-2,1) \in S_{\lambda_2} \text{ with } k \in \mathbb{R}.$$

We can then move between the algebraic definition of eigenvalue, its geometric interpretation, and invariance as a subspace, as shown in **Figure 7**. To exemplify the

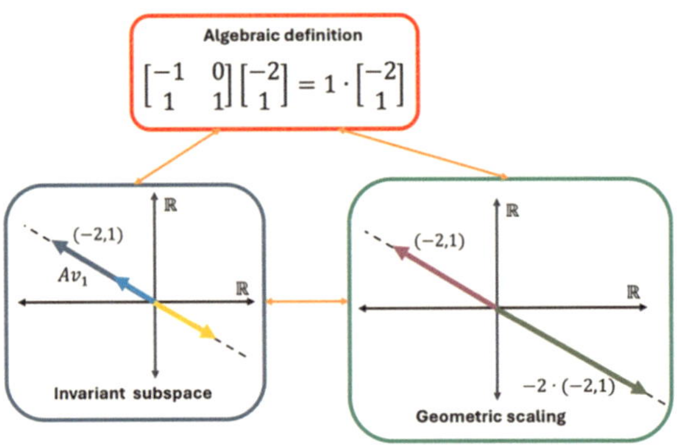

Figure 7.
Interaction of eigenvalue $\lambda_2 = -1$, with geometric scaling (dilatation) and the associated invariant subspace.

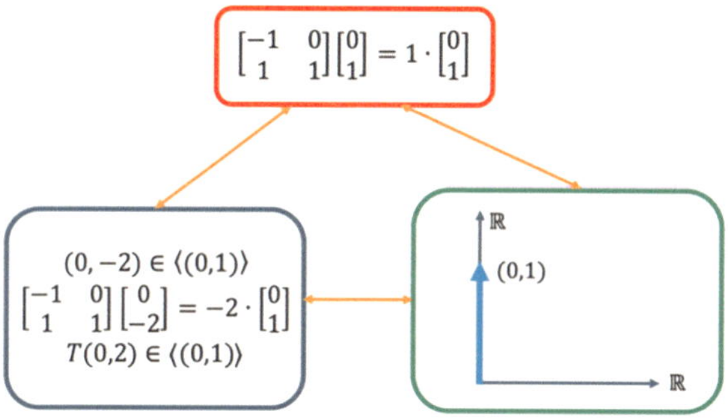

Figure 8.
Interaction of eigenvalue $\lambda_1 = 1$, with geometric scaling and associated invariant subspace.

latter, consider the vector $v_1 = \left(1, \frac{-1}{2}\right) \in \langle(-2,1)\rangle$, $v_1 = \left(1, \frac{-1}{2}\right) \in \langle(-2,1)\rangle$, because $\begin{pmatrix}-1 & 0\\ 1 & 1\end{pmatrix}\begin{pmatrix}1\\ -\frac{1}{2}\end{pmatrix} = \begin{pmatrix}-1\\ \frac{1}{2}\end{pmatrix} = \frac{1}{2}\begin{pmatrix}-2\\ 1\end{pmatrix} \in \langle(-2,1)\rangle.$

Second: The vector $(0,1)$ defines a direction, in this case, coinciding with the Y axis. Any other vector multiple of it remains in the same direction under the linear transformation, which coincides with the fact that it defines an invariant subspace of dimension 1. This direction is determined by the subspace $\langle(0,1)\rangle$ and geometrically corresponds to a straight line, as shown in **Figure 8**.

6. Application of the eigenvalue

It is then presented in $\mathbb{R}^2$. An application of eigenvalues interacting with geometric scaling and invariant subspaces is presented in a Structural Vibration problem.

The study of infinitesimal free vibrations of elastic systems is of great interest in classical vibration theory. A model, which has generated much interest in the literature as a prototype for vibrating structures, is a thin rod of length L (**Figure 9**) with longitudinal vibration governed by the equation

$$\frac{\partial}{\partial x}\left(EA\frac{\partial U}{\partial x}\right) = \rho A\frac{\partial^2 U}{\partial t^2}, 0 < x < L, t > 0 \quad (2)$$

with fixed-free end conditions $u(0) = 0 = u'(L)$. Here, $A = A(x), E = E(x)$, and $\rho = \rho(x)$ are the cross-section area, Young's modulus, and mass density per unit length, respectively.

It is well known that for free vibration of frequency ω, the longitudinal displacement $u(x,t)$ can be written as $u(x,t) = u(x)\sin(\omega t)$, where $u = u(x)$ satisfies the eigenvalue equation

Figure 9.
Thin rod.

$$\frac{d}{dx}\left(EA\frac{du}{dx}\right)+\lambda\rho Au=0, 0<x<L \tag{3}$$

where $\lambda=\omega^2$. For convenience, we will assume that the rod is uniform with fixed-free end conditions, i.e., attached left-hand and free right-hand end.

To better understand the phenomenon of bar vibration, we will apply the finite difference technique in Eq. (3). This will help to better understand the phenomenon since it will be analyzed in small parts of the bar. To do this, let us consider the domain $\Omega=(0,\mathrm{L})$, which represents the possible universe of points into which the rod will be subdivided.

Consider the following partition into n points of Ω:

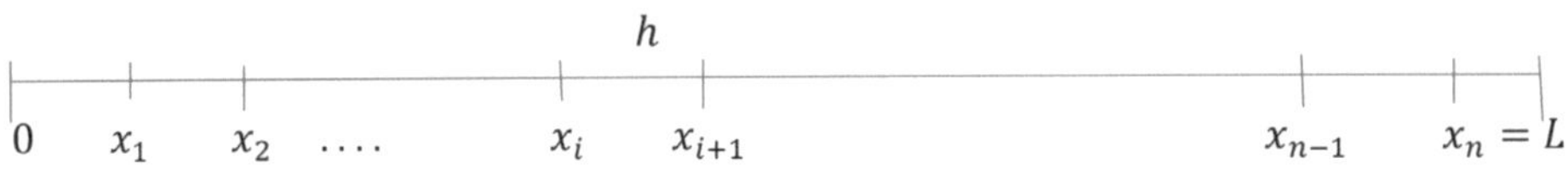

where $h=\frac{1}{n}=x_{i+1}-x_i$, $0\leq i\leq n-1$. Since we are going to look at the continuous Eq. (3) in detail, we need to consider its terms in the same way. That is why physical quantities $A=A(x), E=E(x)$, and $\rho=\rho(x)$ must be evaluated point to point: $E_i=E(x_i), \rho_i=\rho(x_i), A_i=A(x_i), u_i=u(u_i)$.

Since Eq. (3) contains derivatives, we will use a second-order approximation for the second derivative:

$$f''(x_i)\approx\frac{f(x_{i+1})-2f(x_i)+f(x_{i-1})}{h^2} \tag{4}$$

From: $\frac{d}{dx}\left(EA\frac{du}{dx}\right)\approx\frac{E_{i+1}A_{i+1}(u_{i+1}-u_i)-E_iA_i(u_i-u_{i-1})}{h^2}$, we obtain the discrete version of (3): $\frac{E_{i+1}A_{i+1}(u_{i+1}-u_i)-E_iA_i(u_i-u_{i-1})}{h^2}+\lambda\rho_iA_iu_i=0, i=1, \dots, n.$

By multiplying by h and rearranging, we obtain:

$$-k_iu_{i-1}+(k_i+k_{i+1})u_i-k_{i+1}u_{i+1}-\lambda m_iu_i=0, i=1, \dots, n \tag{5}$$

where $k_i=\frac{E_iA_i}{h}$ and $m_i=\rho_iA_ih$ are called stiffness and mass parameters. The stiffness parameter containing E_iA_i is related to elastic properties of the rod. While the mass parameter containing ρ_iA_i is related to density by area unity. The stiffness matrix is given by the symmetric tridiagonal positive definite matrix:

$$\mathrm{K}=\begin{pmatrix} & k_1+k_2 & -k_2 & \cdots & 0 & 0 \\ & -k_2 & k_2+k_3 & -k_3 & \vdots & 0 \\ \ddots & & \ddots & \ddots & \ddots & \ddots \\ 0 & & 0 & -k_{n-1} & k_{n-1}+k_n & -k_n \\ 0 & & 0 & \cdots & -k_n & k_n \end{pmatrix}$$

containing the stiffness parameters and the diagonal positive definite matrix containing the mass parameter is $M=diag\{m_1, \dots, m_n\}$. The boundary conditions $u_0=0=u_{n+1}-u_n$ imply that Eq. (5) can be written as:

$$(K-\lambda M)u=0 \tag{6}$$

where $u=(u_1, \dots \dots, u_n)^t$. Eq. (6) is called generalized eigenvalue problem, the eigenvalue $\lambda=\omega^2$ is known by natural frequency and the eigenvector is known by

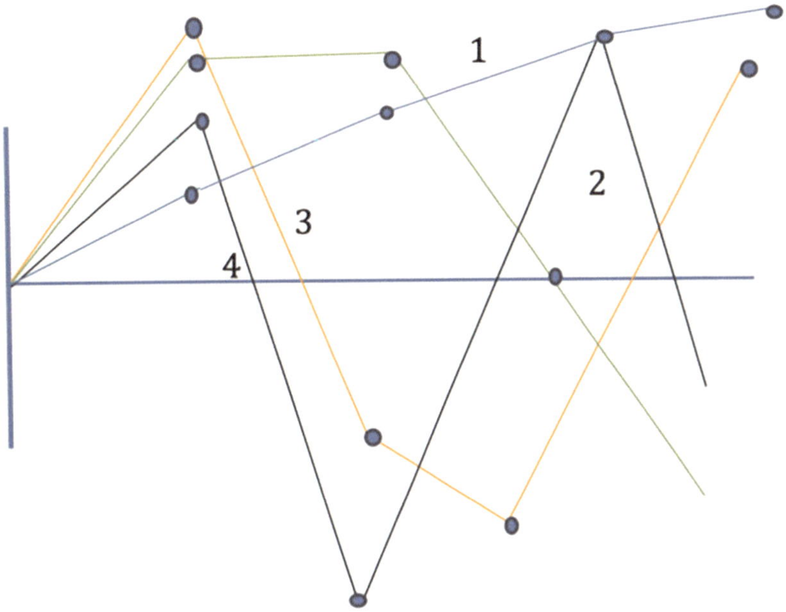

Figure 10.
Vibration modes [1].

vibration mode. For $n = 4$, **Figure 10** shows the vibration modes satisfying the jth modes cross the axis (j−1) times.

We observe from **Figure 10** that:

Mode 1: crosses 0 times axis

Mode 2: crosses 1 time axis

Mode 3: crosses 2 times axis

Mode 4: crosses 3 times axis

Eq. (6) provides important information about the discrete model related to the thin rod. Indeed, in **Figure 11**, a fixed-free spring-mass system is proposed to represent the thin rod. The eigenvector u is related to the displacements about the m_i mass and the eigenvalue λ is related to the vibration frequencies. Natural frequencies are

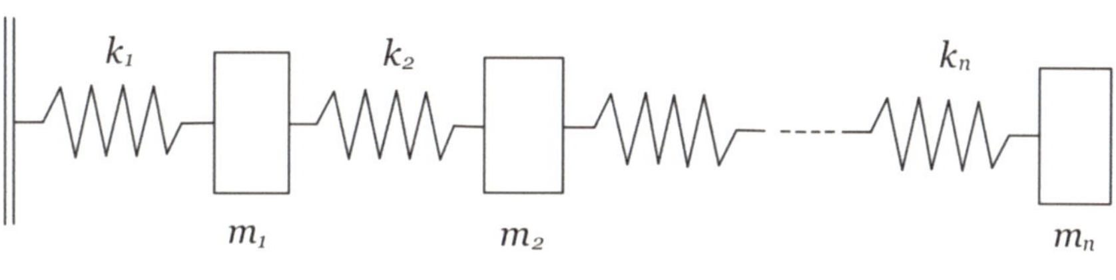

Figure 11.
Spring-mass system [1].

related to vibration modes through the product λu, contracting for $\lambda < 1$ vibration amplitude of masses. For $\lambda > 1$, the vibration oscillations are expanded.

Using the Cholesky decomposition of M, let the B matrix be defined by $B = B^t = diag\{\sqrt{m_1}, \sqrt{m_2}, \ldots, \sqrt{m_n}\}$, satisfying $M = BB^t$ and setting $v^{(i)} = B^t u^{(i)}$, $u^{(i)}$ the ith eigenvector. Multiplying Eq. (6) by B^{-1}, we can transform generalized eigenvalue problem to an equivalent simple eigenvalue problem

$$(J - \lambda I)v = 0 \tag{7}$$

where $J = M^{-1/2}KM^{-1/2}$ belongs to a special matrix called Jacobi matrix, which is symmetric tridiagonal positive definite matrix, negative co-diagonal element. Jacobi matrices have different positive eigenvalues and the eigenvector corresponding to the λ_i eigenvalue in ascending order has $i - 1$ change of signs. Rewriting the eigenvalue Eq. (7) in the form:

$$Jv = \lambda v. \tag{8}$$

From Eq. (8), we can establish the linear transformation $J : \mathrm{V} \to \mathrm{V}$, which transforms an eigenvector v into a multiple vector λv with λ associated eigenvalue. Applying *Theorem 1*, we have that the eigenspace $W = \langle v \rangle$ is J-invariant and then $J(v) = \lambda v \in W$.

Example 2. Let us consider the spring-mass fixed-free vibrating system

Let us consider the spring-mass fixed-free vibrating system in **Figure 12**. The stiffness and mass matrices are given by $K = \begin{pmatrix} k_1+k_2 & -k_2 \\ -k_2 & k_2 \end{pmatrix} = \begin{pmatrix} 1.3 & -0.7 \\ -0.7 & 0.7 \end{pmatrix}$ and $M = \begin{pmatrix} m_1 & 0 \\ 0 & m_2 \end{pmatrix} = \begin{pmatrix} 1.5 & 0 \\ 0 & 2.3 \end{pmatrix}$.

Eq. (6) is now written by $\left(\begin{pmatrix} 1.3 & -0.7 \\ -0.7 & 0.7 \end{pmatrix} - \lambda \begin{pmatrix} 1.5 & 0 \\ 0 & 2.3 \end{pmatrix}\right) v = 0$.

And the equivalent equation with Jacobi matrix $J = \begin{pmatrix} 0.8667 & -03769 \\ -0.3769 & 0.3043 \end{pmatrix}$ is $\left(\begin{pmatrix} 0.8667 & -03769 \\ -0.3769 & 0.3043 \end{pmatrix} - \lambda \begin{pmatrix} 1.0 & 0 \\ 0 & 1.0 \end{pmatrix}\right) v = 0$. Hence, the eigenvalue equation is $\begin{pmatrix} 0.8667 & -03769 \\ -0.3769 & 0.3043 \end{pmatrix} v = \lambda v$.

Computing the eigenvalues using MATLAB function eig(K, M) or eig(J), we have: $\lambda_1 = 0.1153$ and $\lambda_2 = 1.0557$. The corresponding eigenvectors are $v_1 = (-0.4483, -0.8939)$ and $v_2 = (-0.8939, 0.4483)$ and we have equalities:

$$\begin{pmatrix} 0.8667 & -03769 \\ -0.3769 & 0.3043 \end{pmatrix} \begin{pmatrix} -0.4483 \\ -0.8939 \end{pmatrix} = 0.1153 \begin{pmatrix} -0.4483 \\ -0.8939 \end{pmatrix}$$

$$\begin{pmatrix} 0.8667 & -03769 \\ -0.3769 & 0.3043 \end{pmatrix} \begin{pmatrix} -0.8939 \\ 0.4483 \end{pmatrix} = 1.0557 \begin{pmatrix} -0.8939 \\ 0.4483 \end{pmatrix}.$$

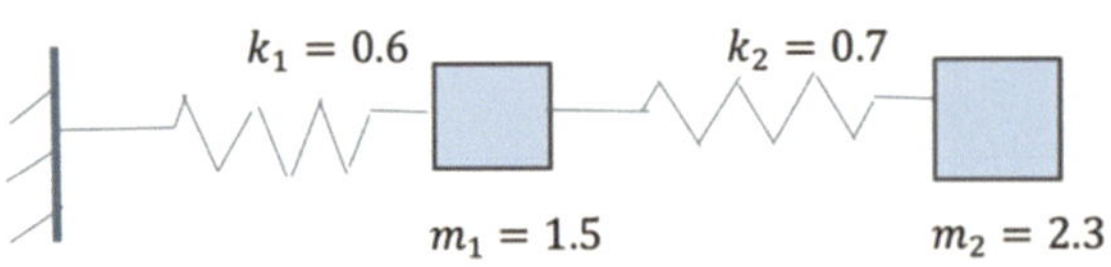

Figure 12.
The spring-mass fixed-free vibrating system.

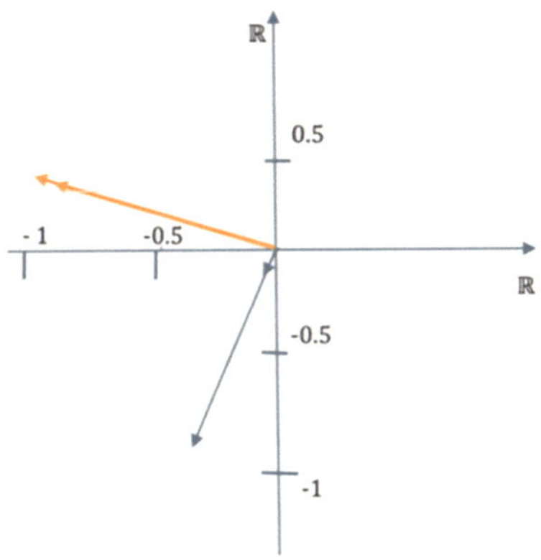

Figure 13.
Scaling of eigenmodes.

We can establish the linear transformation $J : \mathbb{R}^2 \rightarrow \mathbb{R}^2$ and write eigenvalue's equations $J(v_1) = \lambda_1 v_1$ and $J(v_2) = \lambda_2 v_2$. So, in this case, the vector v_1 is contracted by J in $\lambda_1 = 0.1153 < 1$ and the vector v_2 is expanded in $\lambda_2 = 1.0557 > 1$. **Figure 13** shows this fact. Furthermore, $J(v_1) \in \langle v_1 \rangle$ and $J(v_2) \in \langle v_2 \rangle$ because both, $\langle v_1 \rangle$ and $\langle v_2 \rangle$, are invariant subspaces.

We see the scaling of the natural frequencies over the eigenmodes and that they maintain the initial direction. This is an interesting geometric property about natural frequencies and vibration modes. Furthermore, each eigenmode v defines an invariant subspace defined by $W = \langle v \rangle$. Additionally, for $\lambda_1 = 0.1153$, the oscillations will produce a contracted vibration mode, otherwise, $\lambda_2 = 1.0557 > 1$ essentially maintains the vibration oscillation modes.

7. Discussion and conclusions

Linear transformations play a very important role in Linear Algebra and its applications. Within mathematics, they allow us to observe phenomena associated with the transition between vector spaces and how these affect their vectors when they are transformed. In the particular case of endomorphisms (transformations of a vector space itself), the most common way to study their behaviors is to do it locally; that is, their behaviors in subspaces of vector space. This is possible by choosing an arbitrary vector in subspace and seeing how this vector is transformed from its original nature. One way to measure this alteration is through the concept of eigenvalue, which naturally appears as a factor scale of the alteration. It is valid to ask whether the chosen vector, once scaled by the eigenvalue, remains in the subspace and whether this is related to the concept of invariant subspace. Indeed, from the definition of eigenvalue, the scaling of a vector subspace must, by definition, belong to the subspace. This is due to the definition of subspace, which among its properties establishes that the multiples of a vector of the subspace remain in the subspace.

We have then shown that the concept of eigenvalue is related to the concept of scaling and invariant subspaces.

Exploring one-dimensional invariant subspaces within a vector space, using a linear transformation, provides an alternative approach to addressing the concept of eigenvalue. This approach, which prioritizes subspaces before defining eigenvalues as roots of a characteristic polynomial and before identifying eigenvectors, facilitates a deeper understanding of how various concepts of Linear Algebra—such as bases, scaling, linear transformations, vector spaces, and subspaces—are interconnected. In addition, this approach enriches the appreciation of the theoretical structure of Linear Algebra by examining how the geometric scaling of a transformation integrally affects across invariant subspaces. In fact, eigenvectors are recognized as vectors of $\mathbb{R}^2$ *privileged direction*, because they retain their direction and meaning, under the action of an endomorphism in $\mathbb{R}^2$.

As a result of the interaction between eigenvalue, geometric scaling, and invariant subspace in $\mathbb{R}^2$, the following conclusion is established.

When we consider an endomorphism in $\mathbb{R}^2$ with an associated matrix A, each eigenvalue λ has at least one eigenvector v associated with it. The set of all scalar multiples of an eigenvector v, given by $\{\alpha v : \alpha \in \mathbb{R}\}$, defines a one-dimensional invariant subspace. This subspace is invariant because applying A to any vector in this subspace simply scales it by λ, and its result remains in the subspace.

Geometrically in $\mathbb{R}^2$, these invariant subspaces are visualized as lines through the origin. Each line is the direction of expansion or contraction, and the amount of expansion or contraction is given by the eigenvalue λ.

In $\mathbb{R}^2$, the one-dimensional invariant subspaces associated with a matrix are the directions in which the vectors are simply scaled by the eigenvalues of the matrix by linear transformation, without changing their orientations in $\mathbb{R}^2$. These subspaces and eigenvalues offer a deep understanding of how transformation affects vector space $\mathbb{R}^2$.

Finally, the conceptual triad presented in the chapter—*eigenvalues, geometric scaling, and invariant subspace*—has been applied to structural vibration problems in the context of $\mathbb{R}^2$, which can be extended to other applications in structural dynamics but it is necessary to reinterpret the conceptual triad.

Moreover, we consider from the perspective of teaching and learning eigenvalues in $\mathbb{R}^2$ in an undergraduate course, this triad enhances the interaction between relationships and properties that can be visualized through representations, such as those shown in **Table 1**. Specifically, one-dimensional invariant subspaces can be represented as lines passing through the origin and corresponding to sets of eigenvectors associated with an eigenvalue. In this sense, the three concepts of the triad complement each other and provide the students of these topics with an opportunity to “visualize”, which algebraic symbols alone cannot, facilitating reflections on both the concrete and the abstract aspects of Linear Algebra.

Acknowledgements

This work was partially funded by the project PUCV.039.493/2024.

Author details

Marcela Parraguez[1]* and Raúl Jiménez[2]

1 Pontificia Universidad Católica de Valparaíso, Valparaíso, Chile

2 Universidad Católica del Norte, Antofagasta, Chile

*Address all correspondence to: marcela.parraguez@pucv.cl

References

[1] Gladwell GML. Inverse Problems in Vibrations (Solid Mechanics and Its Applications). 2nd edition. Norwell MA. USA: Springer, Kluwer Academic Publishers; 2004

[2] Chu M, Golub G. Inverse Eigenvalue Problems, Theory, Algorithms and Applications. Oxford, England: Oxford University Press; 2005

[3] Gutman I. The Energy of a Graph: Olds and New Results. Algebraic and Combinatorics and Applications. Berlin, Heildelberg: Springer; 2001. DOI: 10.1007/978-3-642-59448-9_13

[4] Varga R. Matrix Iterative Analysis. Melbourne, Victoria, Australia: Hassel Street Press; 2021

[5] Carrel JB. Eigentheory. In: Groups, Matrices, and Vector Spaces. New York, NY: Springer; 2017. DOI: 10.1007/978-0-387-79428-0_8

[6] Wawro M, Watson K, Zandieh M. Student understanding of linear combinations of eigenvectors. In: ZDM. Heidelberg, Germany: Springer Nature; 2018. DOI: 10.1007/s11858-018-01022-8

[7] Lutaif B, Gomez E, Parraguez M, Loureiro G. Linear algebra in engineering: A study of specialized knowledge of Chilean and Brazilian teachers. Teaching Mathematics and Its Applications: An International Journal of the IMA. 2023;**43**:179-203

[8] Betancur A, Roa S, Parraguez M. Construcciones mentales asociadas a los eigenvalores y eigenvectores: refinación de un modelo cognitivo. AIEM—Avances de investigación en educación matemática. 2022;**22**(1):23-46. DOI: 10.35763/aiem22.4005

[9] Parraguez M, Roa-Fuentes S, Jiménez R, Betancur A. Estructuras y mecanismos mentales que desde una perspectiva geométrica modelan y articulan el aprendizaje de valor y vector propio en R2. Revista Latinoamericana de Investigación en Matemática Educativa. 2022;**25**:63-92. DOI: 10.12802/relime.21.2513

[10] Thomas MOJ, Stewart S. Eigenvalues and eigenvectors: Embodied, symbolic and formal thinking. Mathematics Education Research Journal. 2011;**23**(3): 275-296

[11] Salgado H, Trigueros M. Teaching eigenvalues and eigenvectors using models and APOS theory. The Journal of Mathematical Behavior. 2015;**39**:100-120

[12] Krippendorff K. Content Analysis: An Introduction to its Methodology. 4th ed. NY. USA: SAGE Publications; 2018

[13] Mayring P. Qualitative content analysis: Theoretical background and procedures. In: Approaches to Qualitative Research in Mathematics Education. Heidelberg, Germany: Springer; 2015. DOI: 10.1007/978-94-017-9181-6_13

[14] Niss MA. The concept and role of theory in mathematics education: Plenary presentation. In: Relating Practice and Research in Mathematics Education: Proceedings of NORMA 05, Fourth Nordic Conference on Mathematics Education. TAPIR Akademisk Forlag; 2007. Available from: https://rafhladan.is/bitstream/handle/10802/8437/NORMA05.pdf?sequence=1

[15] Hernández-Sampieri R, Mendoza CP. Metodología de la investigación: las rutas cuantitativa,

cualitativa y mixta. México: McGraw-Hill; 2018

[16] Poole D. Álgebra Lineal: Una introducción moderna. 3rd edition. Boston, USA: Cengage Learning Editors; 2011

[17] Howard A. Elementary Linear Algebra. 10th ed. NJ. USA: John Wiley & Sons Inc; 2010

[18] Axler S. Linear Algebra Done Right (Third Edition) Undergraduate Texts in Mathematics. NY. USA: Springer; 2015

Chapter 2

Eigenvalues and Eigenvectors in Controllability Analysis

Raju K. George and Abhijith Ajayakumar

Abstract

This chapter demonstrates the effectiveness of spectral theory in analyzing controllability property of linear and non-linear systems. System design, which is at the core of control system theory, relies heavily on spectral properties of the system matrices. With the aid of eigenvalues and eigenvectors, the state, input, and output matrices can be chosen so that the system behaves in a desired manner. We start by introducing the notion of controllability for linear time-variant (LTI) systems in terms of the spectral properties of the controllability Gramian derived from the state transition matrix of the system. We also define steering control using eigenvalues and eigenvectors of the controllability Gramian, which steers the system from an arbitrary initial state to a desired final state. For LTI systems, the eigenvalues and eigenvectors of the state matrix are used to characterize controllable and observable systems. Key concepts such as the Popov-Belevitch-Hautus(PBH) eigenvector test and Kalman's rank condition are introduced to illustrate how these spectral tools guide the analysis and design of control systems. For nonlinear systems, the spectral properties of the controllability Gramian help in developing a computational algorithm for steering control. This chapter explores how spectral methods help in developing suitable control strategies and give a better understanding of system dynamics and system design.

Keywords: controllability, observability, eigenvalues, eigenvectors, dynamical systems

1. Introduction

Consider a one dimensional system

$$\frac{dx}{dt} = -2x, x(0) = 3 \tag{1}$$

The solution of the system is $x(t) = 3e^{-2t}$ as in **Figure 1**.

Now add a term *sin t* called the forcing term or control term. Then the system is given by

$$\frac{dx}{dt} = -2x + \sin\ t, x(0) = 3 \tag{2}$$

The solution or the trajectory of the system is changed, as in **Figure 2**. The evolution of the system is changed by adding a new forcing term to the system. Thus, the system with a forcing term is called a control system.

The controllability problem is to check the existence of a forcing term or control function $u(t)$ such that the corresponding solution of the system with the initial point $x(t_0) = x_0$ will pass through a desired final point $x(t_1) = x_1$.

Consider a control system characterized by a differential equation of the form

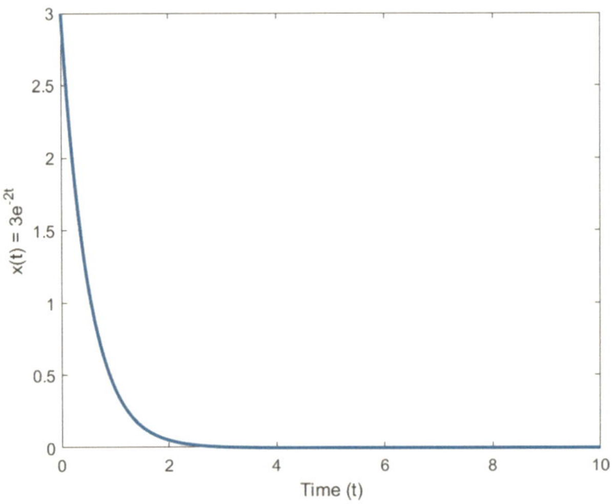

Figure 1.
Solution of the system (1), $x(t) = 3e^{-2t}$. Observe that the system does not has any forcing terms.

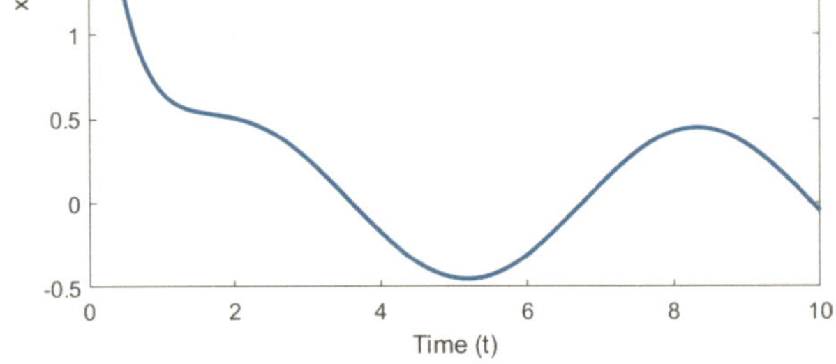

Figure 2.
Solution of the system (2), $x(t) = \frac{1}{5}[2\sin\ t - \cos\ t] + \frac{16}{5}e^{-2t}$. Observe that the solution of the system have changed significantly after the addition of a forcing term.

$$\dot{x}(t) = ax(t) + bu(t), x(t_0) = x_0 \tag{3}$$

where a, b $(b \neq 0)$ are constants. We have to find a control term $u(t)$ so that the solution $x(t)$ passes through $x(t_0) = x_0$ and $x(t_1) = x_1$. Choose a differentiable function $z(t)$ satisfying $z(t_0) = x_0$ and $z(t_1) = x_1$. For example, take

$$z(t) = x_0 + \frac{(x_1 - x_0)}{t_1 - t_0}(t - t_0)$$

Clearly $z(t_0) = x_0$ and $z(t_1) = x_1$.

Define a control using the function z by

$$u = \frac{1}{b}[\dot{z} - az]$$

Then Eq. (3) gives

$$\dot{x} = ax + b\left\{\frac{1}{b}[\dot{z} - az]\right\}$$

That is,

$$\dot{x} - \dot{z} = a(x - z)$$

and $(x - z)(t_0) = 0$. Take $y = (x - z)$. Then the above equation becomes an initial value problem of the form,

$$\frac{dy}{dt} = ay, \;\; y(t_0) = 0$$

As the above system is in variable separable form, we can easily solve it and the unique solution of the system is

$$y(t) = x(t) - z(t) = 0$$

That is, $x(t) = z(t)$ is the solution of the control system satisfying the required condition $x(t_0) = x_0$ and $x(t_1) = x_1$. Note that the control function $u(t)$ not only steers the system from the initial state x_0 to the final state x_1, but also steers the system along the prescribed trajectory $z(t)$. Such a notion is called the *Trajectory Controllability* [1].

Now, consider an n-dimensional dynamical system defined on the time interval $[t_0, t_1]$ characterized by the following equation:

$$\dot{x}(t) = A(t)x(t) + B(t)u(t), \;\; x(t_0) = x_0 \tag{4}$$

where, $x(t) \in \mathbb{R}^n$ is the state vector and $u(t) \in \mathbb{R}^m$ is the control input vector. $A(t) = [a_{ij}(t)] \in \mathbb{R}^{n \times n}$ and $B(t) = [b_{ij}(t)] \in \mathbb{R}^{n \times m}$ are continuous in some interval $[t_0, t_1]$ and are called state matrix and control matrix, respectively. If the state and control matrices of Eq. (2) do not change with time, then the system is called linear time-invariant (LTI) otherwise, it is called a linear time-variant (LTV) system. We will also consider nonlinear systems of the form

$$\dot{x}(t) = A(t)x(t) + B(t)u(t) + f(t, x(t)), \;\; x(t_0) = x_0 \tag{5}$$

where $f(t, x(t))$ is a nonlinear function.

2. Fundamental matrix solution and transition matrix

Let $\{x_0^i : i = 1, 2, \ldots, n\}$ be a basis of $\mathbb{R}^n$. For each i, let $\phi_i(t) \in \mathbb{R}^n$ be the unique solution to the homogeneous system

$$\dot{x}(t) = A(t)x(t) \tag{6}$$

with initial condition $x(t_0) = x_0^i$. Now, $\{\phi_i(t) : i = 1, 2, \ldots, n\}$ is the basis of the solution space of the homogeneous system (4). Consider the $n \times n$ matrix

$$\Phi(t) = [\phi_1(t)\,|\phi_2(t)|\ldots|\phi_n(t)] \tag{7}$$

with n linearly independent solutions of (4) as columns. $\Phi(t)$ is called a *fundamental matrix solution* and it satisfies $\dot{\Phi}(t) = A(t)\Phi(t)$. Clearly, $\Phi(t)$ is non-singular for each t; that is, all eigenvalues of $\Phi(t)$ are positive. It is clear that any matrix $\Phi(t)$ is a fundamental matrix solution to the homogeneous system (4) if and only if $\Phi(t)$ is a solution matrix to the corresponding matrix differential equation $\dot{X}(t) = A(t)X(t)$ and the columns of $\Phi(t)$ are linearly independent. For any non-singular matrix $M \in \mathbb{R}^{n \times n}$, consider the matrix $\Phi(t)M$. We have

$$\frac{d(\Phi(t)M)}{dt} = \dot{\Phi}(t)M = (A(t)\Phi(t))M = A(t)[\Phi(t)M]$$

Also, the columns of $\Phi(t)M$ are linearly independent. Thus, for any non-singular matrix $M \in \mathbb{R}^{n \times n}$, $\Phi(t)M$ is also a fundamental matrix solution of Eq. (4). Then, the *state transition matrix* of the homogeneous system is defined by

$$\Phi(t, t_0) = \Phi(t)\Phi^{-1}(t_0), \quad t_0 \le t \le t_1 < \infty \tag{8}$$

The state transition matrix $\Phi(t, t_0)$ has the following properties:

1. $\Phi(t, t) = I_n, \quad \forall t \in [t_0, \infty)$, where I_n denote the $n \times n$ identity matrix.

2. $\Phi^{-1}(t, t_0) = \Phi(t_0, t)$

3. $\Phi(.,.)$ satisfies the semi-group property

$$\Phi(t, s) = \Phi(t, \tau)\Phi(\tau, s), \quad \forall t_0 \le \tau \le s \le t_1 < \infty$$

4. $\dot{\Phi}(t, t_0) = A(t)\Phi(t, t_0)$

5. $\Phi(t, t_0)$ is the unique solution of the matrix initial value problem

$$\dot{X}(t) = A(t)X(t), \quad X(t_0) = I_n$$

The state transition matrix $\Phi(t, t_0)$ for Eq. (4) is given by the *Peano-Baker series*:

$$\begin{aligned}\Phi(t, t_0) = I_n &+ \int_{t_0}^{t} A(\sigma_1)d\sigma_1 + \int_{t_0}^{t} A(\sigma_1)\int_{t_0}^{\sigma_1} A(\sigma_2)d\sigma_2 d\sigma_1 \\ &+ \int_{t_0}^{t} A(\sigma_1)\int_{t_0}^{\sigma_1} A(\sigma_2)\int_{t_0}^{\sigma_2} A(\sigma_3)d\sigma_3 d\sigma_2 d\sigma_1 + \ldots\end{aligned} \tag{9}$$

This series converges uniformly and absolutely for all $t_0 \leq t \leq t_1 < \infty$. If Eq. (4) is a LTI system, that is, if $A(t) = A$, then the state transition matrix reduces to the matrix exponential given by

$$\Phi(t, t_0) = e^{A(t-t_0)} = I_n + A(t - t_0) + A^2 \frac{(t - t_0)^2}{2!} + A^3 \frac{(t - t_0)^3}{3!} + \ldots \tag{10}$$

For a diagonal matrix A, computing the matrix exponential is straightforward. Let $A = \begin{pmatrix} a_1 & 0 & \cdots & 0 \\ 0 & a_2 & \cdots & 0 \\ \vdots & 0 & \ddots & 0 \\ 0 & 0 & \cdots & a_n \end{pmatrix}$, we have $A^k = \begin{pmatrix} a_1^k & 0 & \cdots & 0 \\ 0 & a_2^k & \cdots & 0 \\ \vdots & 0 & \ddots & 0 \\ 0 & 0 & \cdots & a_n^k \end{pmatrix}$ for $k = 1, 2, \ldots$. Then,

$$e^{A(t-t_0)} = \begin{pmatrix} \sum_{j=0}^{\infty} a_1^j (t - {}_0)^j & 0 & \cdots & 0 \\ 0 & \sum_{j=0}^{\infty} a_2^j (t - t_0)^j & \cdots & 0 \\ \vdots & 0 & \ddots & 0 \\ 0 & 0 & \cdots & \sum_{j=0}^{\infty} a_n^j (t - t_0)^j \end{pmatrix}$$

$$= \begin{pmatrix} e^{a_1(t-t_0)} & 0 & \cdots & 0 \\ 0 & e^{a_2(t-t_0)} & \cdots & 0 \\ \vdots & 0 & \ddots & 0 \\ 0 & 0 & \cdots & e^{a_n(t-t_0)} \end{pmatrix}$$

For some special classes of matrices, the tools of eigenvalues and eigenvectors make the computation of matrix exponential easier. Remember that for two similar matrices A and B such that $A = PBP^{-1}$, the similarity matrix P is the matrix whose columns are the linearly independent eigenvectors of B. Rigorous analysis of eigenvalues and eigenvectors is available in the book "*A Course in Linear Algebra*" by George and Ajayakumar [2].

i. When A is a diagonalizable matrix, that is, if there exists P such that $A = PDP^{-1}$, where $D = \begin{pmatrix} d_1 & 0 & \cdots & 0 \\ 0 & d_2 & \cdots & 0 \\ \vdots & \ddots & \cdots & 0 \\ 0 & \cdots & \cdots & d_n \end{pmatrix}$ is the matrix whose diagonal entries are eigenvalues of A

$$\begin{aligned} e^{A(t-t_0)} &= e^{\left(PDP^{-1}\right)(t-t_0)} \\ &= I + PDP^{-1}(t - t_0) + \left(PDP^{-1}\right)^2 \frac{(t - t_0)^2}{2!} + \left(PDP^{-1}\right)^3 \frac{(t - t_0)^3}{3!} + \cdots \\ &= I + PDP^{-1}(t - t_0) + \left(PD^2P^{-1}\right) \frac{(t - t_0)^2}{2!} + \left(PD^3P^{-1}\right) \frac{(t - t_0)^3}{3!} + \cdots \\ &= P\left(I + D(t - t_0) + \frac{1}{2!} D^2 (t - t_0)^2 + \frac{1}{3!} D^3 (t - t_0)^3 + \cdots\right) P^{-1} \\ &= Pe^{Dt} P^{-1} \end{aligned}$$

ii. Suppose $\lambda_1, \lambda_2, \dots, \lambda_n$ are the eigenvalues of A with a basis of generalized eigenvectors $\{\tilde{v}_1, \tilde{v}_2, \dots, \tilde{v}_n\}$. Let $P = (\tilde{v}_1 \mid \tilde{v}_2 \mid \dots \mid \tilde{v}_n)$. Then, A can be written as the sum of two matrices S and N, where $S = P \begin{pmatrix} \lambda_1 & 0 & \cdots & 0 \\ 0 & \lambda_2 & \cdots & 0 \\ \vdots & \ddots & \cdots & 0 \\ 0 & \cdots & \cdots & \lambda_n \end{pmatrix} P^{-1}$ and N is nilpotent with index of nilpotency $k \leq n$ (This is called the *Jordan-Chevalley decomposition* of a matrix. Refer to Ref. [3] for more details.) Then, by the properties of the matrix exponential,

$$e^{A(t-t_0)} = \left[P \begin{pmatrix} e^{\lambda_1(t-t_0)} & 0 & \cdots & 0 \\ 0 & e^{\lambda_2(t-t_0)} & \cdots & 0 \\ \vdots & \vdots & \ddots & 0 \\ 0 & \cdots & \cdots & e^{\lambda_n(t-t_0)} \end{pmatrix} P^{-1} \right] \left[\sum_{i=0}^{k-1} \frac{1}{i!} N^i (t-t_0)^i \right]$$

If $\Phi(t, t_0)$ is the state transition matrix in Eq. (4) with initial condition $x(t_0) = x_0$, then any future state $x(t)$ can be written by using the state transition matrix $\Phi(t, t_0)$ as

$$x(t) = \Phi(t, t_0)x_0$$

Hence, the name transition matrix for $\Phi(t, t_0)$. Now, a solution to the non-homogeneous system (4) can be obtained by using the transition matrix as follows. Let $\Phi(t, t_0)$ be the transition matrix of the homogeneous system $\dot{x} = A(t)x$. Consider the transformation

$$z(t) = \Phi(t_0, t)x(t)$$

Then

$$x(t) = \Phi(t, t_0)z(t) \tag{11}$$

Differentiating with respect to t,

$$\dot{x}(t) = \dot{\Phi}(t, t_0)z(t) + \Phi(t, t_0)\dot{z}(t)$$

This implies that

$$A(t)x(t) + B(t)u(t) = A(t)\Phi(t, t_0)z(t) + \Phi(t, t_0)\dot{z}(t) = A(t)x(t) + \Phi(t, t_0)\dot{z}(t)$$

Thus, we have

$$B(t)u(t) = \Phi(t, t_0)\dot{z}(t)$$

and as $\Phi^{-1}(t, t_0) = \Phi(t_0, t)$,

$$\dot{z}(t) = \Phi(t_0, t)B(t)u(t)$$

Integrating over t_0 to t,

$$z(t) - z(t_0) = \int_{t_0}^{t} \Phi(t_0, \tau)B(\tau)u(\tau)d\tau$$

which implies,

$$z(t) = z(t_0) + \int_{t_0}^{t} \Phi(t_0, \tau)B(\tau)u(\tau)d\tau$$

Since $z(t_0) = x_0$,

$$z(t) = x_0 + \int_{t_0}^{t} \Phi(t_0, \tau)B(\tau)u(\tau)d\tau$$

Using (11), we have

$$x(t) = \Phi(t, t_0)x_0 + \Phi(t, t_0)\int_{t_0}^{t} \Phi(t_0, \tau)B(\tau)u(\tau)d\tau$$

By using the semi-group property, we have

$$x(t) = \Phi(t, t_0)x_0 + \int_{t_0}^{t} \Phi(t, \tau)B(\tau)u(\tau)d\tau \tag{12}$$

as the required solution to the non-homogeneous system. For a thorough exposition of the fundamental matrix and transition matrix, refer to Ref [4].

3. Controllability of linear systems

Definition 1 (**Controllability**): The system (4) is controllable in a time interval $[t_0, t_1]$ if, given any two states $x_0, x_1 \in \mathbb{R}^n$, there exists an admissible control function $u \in \mathscr{L}^2([t_0, t_1], \mathbb{R}^m)$, such that the corresponding solution of (4) with the initial condition $x(t_0) = x_0$ also passes through the desired final point $x(t_1) = x_1$.

For a concise introduction to controllability, refer to Refs [5, 6].

3.1 Characterization for controllability

From the definition, Eq. (4) is controllable if and only if there exists $u \in \mathscr{L}^2([t_0, t_1], \mathbb{R}^m)$ such that

$$x_1 = x(t_1) = \Phi(t_1, t_0)x_0 + \int_{t_0}^{t_1} \Phi(t_1, \tau)B(\tau)u(\tau)d\tau$$

Then,

$$x_1 - \Phi(t_1, t_0)x_0 = \int_{t_0}^{t_1} \Phi(t_1, \tau)B(\tau)u(\tau)d\tau$$

Denote $x_1 - \Phi(t_1, t_0)x_0 = w$, then

$$w = \int_{t_0}^{t_1} \Phi(t_1, \tau)B(\tau)u(\tau)d\tau \tag{13}$$

Thus, the system (4) is controllable if and only if for every $w \in \mathbb{R}^n$, there exists $u \in \mathcal{L}^2([t_0, t_1], \mathbb{R}^m)$ such that Eq. (13) is satisfied. Define an operator $\mathcal{C}$: $\mathcal{L}^2([t_0, t_1]; \mathbb{R}^m) \to \mathbb{R}^n$ by

$$\mathcal{C}u = \int_{t_0}^{t_1} \Phi(t_1, \tau) B(\tau) u(\tau) d\tau \tag{14}$$

Thus, the system (4) is controllable if and only if the operator $\mathcal{C}$ is onto. Obviously, $\mathcal{C}$ is a bounded linear operator and $\mathcal{C}$ defines its adjoint operator $\mathcal{C}^* : \mathbb{R}^n \to \mathcal{L}^2([t_0, t_1], \mathbb{R}^m)$ in the following way:

$$\begin{aligned}
\langle \mathcal{C}^* v, u \rangle_{\mathcal{L}^2} &= \langle v, \mathcal{C}u \rangle_{\mathbb{R}^n}, \forall \ u \in \mathcal{L}^2([t_0, t_1], \mathbb{R}^m), v \in \mathbb{R}^n \\
&= \left\langle v, \int_{t_0}^{t_1} \Phi(t_1, \tau) B(\tau) u(\tau) d\tau \right\rangle_{\mathbb{R}^n} \\
&= \int_{t_0}^{t_1} \langle v, \Phi(t_1, \tau) B(\tau) u(\tau) \rangle_{\mathbb{R}^n} d\tau \\
&= \int_{t_0}^{t_1} \left\langle B^T(\tau) \Phi^T(t_1, \tau) v, u(\tau) \right\rangle_{\mathbb{R}^m} d\tau \\
&= \left\langle B^T(.) \Phi^T(t_1, .) v, u \right\rangle_{\mathcal{L}^2}
\end{aligned}$$

Hence, the adjoint operator of $\mathcal{C}$ is the linear operator $\mathcal{C}^* : \mathbb{R}^n \to \mathcal{L}^2([t_0, t_1]; \mathbb{R}^m)$, given by

$$(\mathcal{C}^* v)(t) = B^T(t) \Phi^T(t_1, t) v \tag{15}$$

The composition of $\mathcal{C}$ and $\mathcal{C}^*$ defines a bounded linear operator $\mathcal{C}\mathcal{C}^* : \mathbb{R}^n \to \mathbb{R}^n$ by,

$$\mathcal{C}\mathcal{C}^* v = \int_{t_0}^{t_1} \Phi(t_1, \tau) B(\tau) B^T(\tau) \Phi^T(t_1, \tau) v d\tau \tag{16}$$

Clearly, the operator $\mathcal{C}\mathcal{C}^*$ can be realized as an $n \times n$ positive semi-definite matrix, called *Controllability Gramian* of the system (4) and is denoted by $\mathcal{W}(t_0, t_1)$. The following theorem relates controllability of (4) and the properties of linear operators $\mathcal{C}, \mathcal{C}^*$ and $\mathcal{C}\mathcal{C}^*$.

Theorem 1: The following statements are equivalent:

i. The system (4) is controllable.

ii. The operator $\mathcal{C}$ is onto.

iii. The adjoint operator $\mathcal{C}^*$ is one-one.

iv. Zero is not an eigenvalue of the Controllability Gramian $\mathcal{W}(t_0, t_1) = \mathcal{C}\mathcal{C}^*$.

Proof: Clearly, $(i) \Leftrightarrow (ii)$ by definition of the operator $\mathcal{C}$ in Eq. (14).

Now, let us show $(ii) \Rightarrow (iii)$. Suppose that $\mathcal{C}$ is onto. We have to show that $\mathcal{C}^*$ is one-one. It is enough to show that $\mathcal{C}^* v = 0$ if and only if $v = 0$. Let $v \in \mathbb{R}^n$ such that $\mathcal{C}^* v = 0$. As $\mathcal{C}$ is onto, there exists $u \in \mathcal{L}^2([t_0, t_1]; \mathbb{R}^m)$ such that $\mathcal{C}u = v$. Then

$$\langle v, v\rangle = \langle Cu, v\rangle = \langle u, C^* v\rangle = \langle u, 0\rangle = 0$$

This implies that $v = 0$. Hence C^* is one-one.

To prove $(iii) \Rightarrow (iv)$, suppose λ is an eigenvalue of CC^*, that is, there exists $v \neq 0$ such that $CC^* v = \lambda v$. Now,

$$\|C^* v\|^2 = \langle C^* v, C^* v\rangle = \langle CC^* v, v\rangle = \langle \lambda v, v\rangle = \|\lambda v\|^2 \geq 0$$

Now, if $\lambda = 0$, then $\|C^* v\|^2 = 0$, which implies that $C^* v = 0$, which is a contradiction to the one-oneness of C^*. Hence $\lambda > 0$.

$(iv) \Rightarrow (i)$ Suppose that zero is not an eigenvalue of $CC^* = \mathcal{W}(t_0, t_1)$, this implies the invertibility of $CC^* = \mathcal{W}(t_0, t_1)$.

Define a control function

$$u(t) = B^T(t)\Phi^T(t_1,)\mathcal{W}^{-1}(t_0, t_1)[x_1 - \Phi(t_1, t_0)x_0] \tag{17}$$

Using this control, the state of the system is given by

$$x(t) = \Phi(t, t_0)x_0 + \int_{t_0}^{t} \Phi(t, \tau)B(\tau)B^T(\tau)\Phi^T(t_1, \tau)\mathcal{W}^{-1}(t_0, t_1)[x_1 - \Phi(t_1, t_0)x_0]d\tau$$

Then,

$$x(t_0) = \Phi(t_0, t_0)x_0 + \int_{t_0}^{t_0} \Phi(t_0, \tau)B(\tau)B^T(\tau)\Phi^T(t_1, \tau)\mathcal{W}^{-1}(t_0, t_1)[x_1 - \Phi(t_1, t_0)x_0]d\tau = x_0$$

and

$$\begin{aligned} x(t_1) &= \Phi(t_1, t_0)x_0 + \int_{t_0}^{t_1} \Phi(t_1, \tau)B(\tau)B^T(\tau)\Phi^T(t_1, \tau)\mathcal{W}^{-1}(t_0, t_1)[x_1 - \Phi(t_1, t_0)x_0]d\tau \\ &= \Phi(t_1, t_0)x_0 + \mathcal{W}(t_0, t_1)\mathcal{W}^{-1}(t_0, t_1)[x_1 - \Phi(t_1, t_0)x_0] = \Phi(t_1, t_0)x_0 + x_1 - \Phi(t_1, t_0)x_0 = x_1 \end{aligned}$$

Since, x_0 and x_1 are arbitrary, the system is controllable.

Spectral analysis has much more to do with controllability analysis of linear - time-invariant systems, that is, when $A(t) = A$ and $B(t) = B$ are constant matrices. The *Popov-Belevitch-Hautus* (PBH) condition gives a relation between the left-eigenspace of the state matrix A and the column space of the input matrix B.

Definition 2 (**Left eigenvector**): Let A be an $n \times n$ matrix. A non-zero row vector v is said to be a left eigenvector of A if $vA = \lambda v$, for some scalar λ.

Note that v is a left eigenvector of A if and only if v^T is a right eigenvector of A^T.

Theorem 2: If the system (4) is LTI, then it is controllable if and only if for every left eigenvector v of A, $v^T B \neq 0$.

Proof: Suppose that system (4) is not controllable. That is, $rank[\mathcal{Q}(A, B)] = r < n$. By Kalman controllability decomposition [7], there exists a non-singular matrix T such that

$$T^{-1}AT = \begin{bmatrix} A_{11} & A_{12} \\ 0 & A_{22} \end{bmatrix} = \hat{A} \quad \text{and} \quad T^{-1}B = \begin{bmatrix} B_1 \\ 0 \end{bmatrix}$$

Let $\tilde{v}$ be an eigenvector of A_{22}^T corresponding to the eigenvalue λ. That is $A_{22}^T\tilde{v} = \lambda\tilde{v}$. This implies that $\tilde{v}^T A_{22} = \overline{\lambda}\tilde{v}^T$. As A_{22} is a real matrix both λ and $\overline{\lambda}$ are eigenvalues of A_{22} and because of the similarity of A and $\hat{A}$ both λ and $\overline{\lambda}$ are eigenvalues of A also. Now, define $v^T = \begin{bmatrix} 0_{1\times r} & \tilde{v}^T \end{bmatrix} T^{-1}$. Then,

$$\begin{aligned} v^T A &= v^T T \begin{pmatrix} A_{11} & A_{12} \\ 0 & A_{22} \end{pmatrix} T^{-1} \\ &= \begin{pmatrix} 0_{1\times r} & \tilde{v}^T \end{pmatrix} \begin{pmatrix} A_{11} & A_{12} \\ 0 & A_{22} \end{pmatrix} T^{-1} \\ &= \begin{pmatrix} 0 & \tilde{v}^T A_{22} \end{pmatrix} T^{-1} \\ &= \begin{pmatrix} 0 & \lambda\tilde{v}^T \end{pmatrix} T^{-1} = \lambda v^T \end{aligned}$$

Also, $v^T B = \begin{pmatrix} 0_{1\times r} & \tilde{v}^T \end{pmatrix} T^{-1} T \begin{pmatrix} B_1 \\ 0 \end{pmatrix} T^{-1} = 0$

To prove the converse part, we need the notion of *Controllability Matrix*. The matrix $Q(A,B) = \begin{bmatrix} B|AB|\cdots|A^{n-1}B \end{bmatrix}$ is said to be the controllability matrix of the LTI system (A, B). (A, B) is controllable if and only if the controllability matrix has full rank. This is the celebrated *Kalman's theorem* (Proof is given in Appendix A). Suppose that there exists $v \neq 0$ such that $v^T A = \lambda A$ and $v^T B = 0$ is satisfied. Then,

$$\begin{aligned} v^T Q(A,B) &= v^T \begin{bmatrix} B|AB|\cdots|A^{n-1}B \end{bmatrix} \\ &= \begin{bmatrix} v^T B|v^T AB|\cdots|v^T A^{n-1}B \end{bmatrix} = \begin{bmatrix} 0|\lambda v^T B|\cdots|\lambda^{n-1}v^T B \end{bmatrix} = 0 \end{aligned}$$

which implies that $rank[Q(A,B)] < n$. Hence, the system (4) is not controllable.

Note 1: The PBH condition states that an LTI system (A, B) is uncontrollable if and only if there exists a left eigenvector of A that is simultaneously orthogonal to all columns of B. This helps one identify the possible choices of input matrices B such that the system (A, B) is controllable.

The following example establishes the significance of eigenvalues and eigenvectors in system design and in establishing its controllability.

Example 1: Consider a damped harmonic oscillator, given by
$\begin{bmatrix} \dot{x}_1 \\ \dot{x}_2 \end{bmatrix} = \begin{pmatrix} 0 & 1 \\ -\frac{k}{m} & -\frac{d}{m} \end{pmatrix} \begin{pmatrix} x_1 \\ x_2 \end{pmatrix} + \begin{pmatrix} b_1 \\ b_2 \end{pmatrix} u$, where k is the spring constant, m is the mass and d is the damping coefficient. Suppose that we want a critically damped system, which provides a fast response without overshoot. For a critically damped system, the eigenvalues are real and equal. Critical damping occurs when the damping ratio, $\zeta = \frac{Re(\lambda)}{|\lambda|} = 1$, where λ is an eigenvalue. We can choose a desired natural frequency ω_n for the system. Then, the eigenvalues for critical damping are $\lambda_1 = \lambda_2 = -\omega_n$. Then, we get $k = m\omega_n^2$ and $d = 2m\omega_n$. If we want a natural frequency of 5 rad/s and a mass of 1 kg, then $k = 25N/m$ and $d = 10Ns/m$. That is, $A = \begin{bmatrix} 0 & 1 \\ -25 & -10 \end{bmatrix}$. The left eigenspace of A is $E_1 = \{(5v, v)|v \in \mathbb{R}\}$. The orthogonal complement of E_1 is $F_1 = \{(-v, 5v)|v \in \mathbb{R}\}$. Thus, for any choice of B not in F_1, the given system will be

controllable. For instance, if $B = \begin{pmatrix} 5 \\ 1 \end{pmatrix} \notin F_1$, the system (A, B) is controllable and for $B = \begin{pmatrix} -1 \\ 5 \end{pmatrix} \in F_1$, the system (A, B) is not controllable.

In the following example, given a state matrix, we will see how to find possible input matrices such that the system is controllable, with the help of the PBH eigenvector test.

Example 2: Consider a linear system $\dot{x} = Ax + Bu$, where $A = \begin{pmatrix} 1 & 0 \\ 1 & 2 \end{pmatrix}$ and $B = \begin{pmatrix} b_1 \\ b_2 \end{pmatrix}$. The eigenvalues of A are 1, 2 and the corresponding left-eigenspaces are $E_1 = \{(v \quad 0)|v \in \mathbb{R}\}$ and $E_2 = \{(v \quad v)|v \in \mathbb{R}\}$. If we have to choose a B such that controllability is guaranteed, B should not be orthogonal to the left eigenvectors corresponding to 1 and 2. We can easily find that the orthogonal complement of E_1 is $F_1 = \{(0 \quad w)|w \in \mathbb{R}\}$ and that of E_2 is $F_2 = \{(w \quad -w)|w \in \mathbb{R}\}$. Thus, for any choice of B not in F_1 and F_2, the system (A, B) is controllable. As a particular instance, for $B = \begin{pmatrix} 1 \\ 2 \end{pmatrix} \notin F_1, F_2$, the system (A, B) is controllable and for $B = \begin{pmatrix} 0 \\ 1 \end{pmatrix} \in F_1$, (A, B) is uncontrollable.

The next theorem gives an explicit expression for steering control in terms of eigenvectors of the controllability Gramian matrix.

Theorem 3 [8]: For the control system (4), the control function defined as

$$u(t) = B^T(t)\Phi^T(t_1, t)\sum_i \frac{c_i v_i}{\lambda_i} \tag{18}$$

steers the system from x_0 to x_1, where λ_i is the ith eigenvalue of the Gramian matrix $\mathcal{W}(t_0, t_1)$, $\{v_n\}$ is the orthonormal basis of $\mathbb{R}^n$ generated by eigenvectors corresponding to $\{\lambda_i\}_{i=1}^n$ and c_i's are the coordinates of the vector $x_1 - \Phi(t_1, t_0)x_0$ with respect to $\{v_n\}$.

Proof: Since $\mathcal{W}(t_0, t_1)$ is a symmetric matrix, it has n linearly independent eigenvectors $v_1, v_2, \dots, v_n$ forming an orthonormal basis of $\mathbb{R}^n$. Consider the vector $x_1 - \Phi(t_1, t_0)x_0$. As $\{v_i | i = 1, 2, \dots, n\}$ forms a basis, there exist scalars $c_i, i = 1, 2, \dots, n$ such that

$$x_1 - \Phi(t_1, t_0)x_0 = \sum_{i=1}^{n} c_i v_i \tag{19}$$

is the unique representation of $x_1 - \Phi(t_1, t_0)x_0$ with respect to the given basis. Now, we claim that the control defined by Eq. (18), steers the system (4) from x_0 to x_1 during the time $[t_0, t_1]$. From equation, we have

$$x(t_0) = \Phi(t_0, t_0)x_0 + \int_{t_0}^{t_0} \Phi(t_0, s)B(s)u(s)ds = x_0$$

and at $t = t_1$,

$$x(t_1) = \Phi(t_1, t_0)x_0 + \int_{t_0}^{t_1} \Phi(t_1, s)B(s)u(s)ds$$

using Eq. (18), we have

$$x(t_1) = \Phi(t_1,t_0)x_0 + \int_{t_0}^{t_1} \Phi(t_1,s)B(s)B^T(s)\Phi^T(t_1,s)\sum_{i=1}^{n}\frac{c_i}{v_i}\lambda_i ds$$

$$= \Phi(t_1,t_0)x_0 + \sum_{i=1}^{n}\frac{c_i}{\lambda_i}\int_{t_0}^{t_1}\Phi(t_1,s)B(s)B^T(s)\Phi^T(t_1,s)v_i ds = \Phi(t_1,t_)x_0 + \sum_{i=1}^{n}\frac{c_i}{\lambda_i}\mathcal{W}(t_0,t_1)v_i$$

$$= \Phi(t_1,t_0)x_0 + \sum_{i=1}^{n}\frac{c_i}{\lambda_i}\lambda_i v_i = \Phi(t_1,t_0)x_0 + \sum_{i=1}^{n}c_i v_i = \Phi(t_1,t_0)x_0 + x_1 - \Phi(t_1,t_0)x_0 = x_1$$

Hence, the system is controllable.

Example 3: Consider the damped harmonic oscillator system discussed in Example 1, with $A = \begin{pmatrix} 0 & 1 \\ -25 & -10 \end{pmatrix}$ and $B = \begin{pmatrix} 5 \\ 1 \end{pmatrix}$. Suppose that we need to steer the system from $\begin{pmatrix} 0 \\ 0 \end{pmatrix}$ to $\begin{pmatrix} 1 \\ 1 \end{pmatrix}$. Let us compute the steering control using the above theorem. We can see that the controllability Gramian matrix is, $\mathcal{W}(0,1) = \begin{pmatrix} 6.452 & -12.5 \\ -12.5 & 31.3 \end{pmatrix}$ (MATLAB code is provided in the Appendix part). The eigenvalues are $\lambda_1 = 1.252, \lambda_2 = 36.5$ and the corresponding eigenvectors are $v_1 = \begin{pmatrix} -0.9233 \\ -0.3841 \end{pmatrix}, v_2 = \begin{pmatrix} -0.3841 \\ -0.9233 \end{pmatrix}$, respectively. Then, the control input (**Figure 3**)

$$u(t) = B^T\Phi^T(t_1,t)\sum_i \frac{c_i v_i}{\lambda_i}$$

$$= \frac{848625}{28561}e^{5t-5} + \frac{11845}{28561}e^{5t-5}(5t-4) - \frac{54750}{2197}te^{5t-5} + \frac{1480625}{28561}e^{5t-5}(t-1)$$

Also, from Eq. (12) we can see that the trajectory is

$$\begin{pmatrix} x_1(t) \\ x_2(t) \end{pmatrix} = \begin{pmatrix} -e^{10t-10}\left(\frac{980t^2}{13} - \frac{28190t}{169} + \frac{15281}{169}\right) \\ e^{10t-10}\left(\frac{4900t^2}{13} - \frac{128210t}{169} + \frac{64679}{169}\right) \end{pmatrix}$$

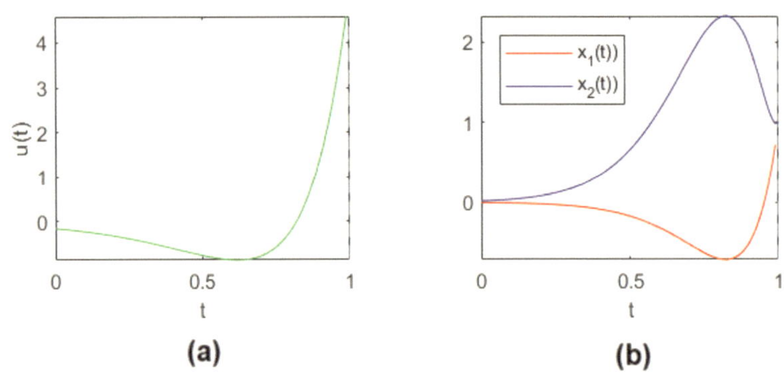

Figure 3.
(a) Control input. (b) Steered state trajectory.

4. Observability of linear systems

Observability is another core notion in control systems theory that focuses on the ability to derive the internal state of a system from its output measurements. A system is considered observable in control theory if its complete state can be uniquely inferred from the given output information. Because an observable system enables for reliable monitoring and assessment of its internal dynamics, it is critical in devising successful control strategies. Consider the system (4) with an output equation as follows:

$$\begin{aligned} \dot{x}(t) &= A(t)x(t) + B(t)u(t) \\ y(t) &= C(t)x(t) \end{aligned} \tag{20}$$

where, $y(t) \in \mathbb{R}^p$ and $C(t) \in \mathbb{R}^{p\times n}$ are the output vector and output matrix, respectively. Now, we have the following formal definition for observability.

Definition 3 (**Observability**): The system (20) is said to be observable over a time period $[t_0, t_1]$ if it is possible to determine uniquely the initial state $x(t_0) = x_0$ from the knowledge of the output $y(t)$ over the time period $[t_0, t_1]$.

Let $\Phi(t, t_0)$ be the state transition matrix of the homogeneous system $\dot{x}(t) = Ax(t)$. The unique solution is given by

$$x(t) = \Phi(t, t_0)x_0$$

Then, the observation equation can be written as

$$y(t) = C(t)x(t) = C(t)\Phi(t, t_0)x_0, \quad t_0 \le t \le t_1$$

4.1 Characterizations for observability

As we have seen in the case for controllability of Eq. (4), we define an operator $\mathscr{M} : \mathbb{R}^n \to \mathscr{L}^2([t_0, t_1]; \mathbb{R}^m)$ by,

$$(\mathscr{M}x_0)(t) = C(t)\Phi(t, t_0)x_0 \tag{21}$$

That is, $(\mathscr{M}x_0)(t) = y(t)$. The initial state is mapped to the observed function. As we need to uniquely determine x_0 from $y(.)$, the system (20) is observable if and only if $\mathscr{M}$ is one-one. Here, the adjoint operator of $\mathscr{M}$ is $\mathscr{M}^* : \mathscr{L}^2([t_0, t_1]; \mathbb{R}^m) \to \mathbb{R}^n$ given by

$$\mathscr{M}^* v = \int_{t_0}^{t_1} \Phi^T(\tau, t_0)C^T(\tau)v(\tau)d\tau \tag{22}$$

The observability Gramian $\mathscr{M}^*\mathscr{M} : \mathbb{R}^n \to \mathbb{R}^n$ is given by

$$\mathscr{M}^*\mathscr{M}v = \mathfrak{M}(t_0, t_1) = \int_{t_0}^{t_1} \Phi^T(\tau, t_0)C^T(\tau)C(\tau)\Phi(\tau, t_0)vdt \tag{23}$$

From any initial state x_0, we have a unique state given by

$$x(t) = \Phi(t, t_0)x_0$$

Thus, observability problem reduces to finding the unique initial state x_0 from the knowledge of y observed on $[t_0, t_1]$. Like Theorem 1, for the controllability of system (4), we have the following theorem for observability of system (20).

Theorem 4: The following statements are equivalent:

i. The system (20) is observable.

ii. The operator $\mathcal{M}$ is one-one.

iii. The adjoint operator $\mathcal{M}^*$ is onto.

iv. Zero is not an eigenvalue of the observability Gramian $\mathfrak{M}(t_0, t_1) = \mathcal{M}^* \mathcal{M}$.

Proof: Proof is similar to that of Theorem 1.

Some kind of interconnections between controllability and observability can be observed. This interconnection is called duality. To delve into the notion of duality, we define the notion of adjoint systems.

4.2 Duality between controllability and observability

The relationship between controllability and observability is dualistic: A system is controllable if and only if its adjoint system is observable. This duality is essential for engineers to understand how manipulation of system inputs can affect the ability to observe its states accurately. This interdependence guides the design of feedback systems and state estimators, ensuring both properties are optimized for better performance. In this section, we will establish the duality between controllability and observability.

Definition 4 (**Adjoint Systems**): A system with state $x(t)$ is said to be adjoint to a system with state $p(t)$ if $\langle x(t), p(t) \rangle$ is a constant. That is, if $\frac{d}{dt} \langle x(t), p(t) \rangle = 0$.

Theorem 5: The systems

$$\dot{x}(t) = A(t)x(t) \tag{24}$$

and

$$\dot{p}(t) = -A^T(t)p(t) \tag{25}$$

are adjoint to each other.

Proof: By the product rule for differentiation concerning inner-product, we have

$$\begin{aligned}
\frac{d}{dt}(\langle x(t), p(t) \rangle) &= \langle \dot{x}(t), p(t) \rangle + \langle x(t), \dot{p}(t) \rangle \\
&= \langle A(t)x(t), p(t) \rangle + \left\langle x(t), -A^T(t)p(t), \right\rangle \\
&= \left\langle x(t), A^T(t)p(t) \right\rangle + \left\langle x(t), -A^T(t)p(t) \right\rangle = \langle x(t), 0 \rangle = 0
\end{aligned}$$

Hence, $\langle x(t), p(t) \rangle$ is a constant, proving that the systems (24) and (25) are adjoint to each other.

The state transition matrices of the above systems are also related as shown in the following theorem.

Theorem 6: If $\Phi(t, t_0)$ is the transition matrix of the system $\dot{x}(t) = A(t)x(t)$, then $\Phi^T(t_0, t)$ is the transition matrix of $\dot{p}(t) = -A^T(t)p(t)$.

Proof: By using the properties of transition matrix, we have $I = \Phi(t, t_0)\Phi(t_0, t)$. Differentiating w.r.t. t,

$$0 = \dot{\Phi}(t, t_0)\Phi(t_0, t) + \Phi(t, t_0)\dot{\Phi}(t_0, t) = A(t)\Phi(t, t_0)\Phi(t_0, t) + \Phi(t, t_0)\dot{\Phi}(t_0, t) = A(t) + \Phi(t, t_0)\dot{\Phi}(t_0, t)$$

This implies that $\Phi(t, t_0)\dot{\Phi}(t_0, t) = -A(t)$ and hence $\dot{\Phi}(t_0, t) = -\Phi(t_0, t)A(t)$. Thus, we have

$$\frac{d\left[\Phi^T(t_0, t)\right]}{dt} = -A^T(t)\Phi^T(t_0, t)$$

Therefore, $\Phi^T(t_0, t)$ satisfies $\dot{p}(t) = -A^T(t)p(t)$. Further, $\Phi^T(t_0, t_0) = I$. Thus, $\Phi^T(t_0, t)$ is the transition matrix to the adjoint system $\dot{p}(t) = -A^T(t)p(t)$.

Theorem 7: Consider the linear control system

$$\dot{x}(t) = A(t)x(t) + B(t)u(t) \tag{26}$$

and the input-free observation system

$$\begin{aligned} \dot{x}(t) &= -A^T(t)x(t) \\ y(t) &= B^T(t)x(t) \end{aligned} \tag{27}$$

System (26) is controllable if and only if adjoint system (27) is observable.

Proof: Suppose that the adjoint system (27) is observable.

$$\begin{aligned} \text{System (27) is observable} \Leftrightarrow \mathfrak{M}(t_0, t_1) &= \int_{t_0}^{t_1} \left[\Phi^T(t_0, \tau)\right]^T \left[B^T(\tau)\right]^T B^T(\tau)\Phi^T(t_0, \tau)d\tau \quad \text{is invertible} \\ &\Leftrightarrow \int_{t_0}^{t_1} \Phi(t_0, \tau)B(\tau)B^T(\tau)\Phi^T(t_0, \tau)d\tau \;\text{ is invertible} \\ &\Leftrightarrow \int_{t_0}^{t_1} \Phi(t_1, t_0)\Phi(t_0, \tau)B(\tau)B^T(\tau)\Phi^T(t_1, t_0)\Phi^T(t_0, \tau)d\tau \;\text{ is invertible} \end{aligned}$$

as both $\Phi(t_1, t_0)$ and $\Phi^T(t_1, t_0)$ are invertible,

$$\begin{aligned} &\Leftrightarrow \int_{t_0}^{t_1} \Phi(t_1, \tau)B(\tau)B^T(\tau)\Phi^T(t_1, \tau)d\tau \;\text{ is invertible} \\ &\Leftrightarrow \mathcal{W}(t_0, t_1) \text{is invertible} \\ &\Leftrightarrow \text{System (26) is controllable} \end{aligned}$$

Thus, system (26) is controllable if and only if adjoint system (27) is observable.

The notion of duality asserts that if a linear system is controllable, it shares similar structural properties with its dual, observable system. This means that the matrices associated with controllability and observability exhibit analogous patterns. Understanding duality is essential in designing balanced and well-behaved control systems, ensuring that controllability and observability are appropriately matched for optimal performance and stability. The concept of duality aids in the translation of similar LTI system conditions from the case of controllability to the case of observability for adjoint systems.

We can also obtain the following PBH observability condition for the LTI system (20) in terms of eigenvalues and eigenvectors.

Theorem 8: If the system (20) is LTI, then it is observable if and only if for every right eigenvector w of A, $Cw \neq 0$.

Proof: Suppose that the system (18) is observable. Then, $(-A^T, C^T)$ is controllable. By Theorem 2, for any left eigenvector w of A, $w^T C^T \neq 0$. Taking conjugate transpose, for any right eigenvector w of A, $Cw \neq 0$. Similarly, the converse follows.

Example 4: Consider the observation system with the same state matrix A as in the previous example and $C = \begin{pmatrix} c_1 & c_2 \end{pmatrix}$. The eigenvalues of A are 1 and 2, with the corresponding right-eigenspaces $E_1 = \left\{ \begin{pmatrix} v \\ -v \end{pmatrix} | v \in \mathbb{R} \right\}$ and $E_2 = \left\{ \begin{pmatrix} 0 \\ v \end{pmatrix} | v \in \mathbb{R} \right\}$ Note that the orthogonal complements of E_1 and E_2 are $F_1 = \left\{ \begin{pmatrix} w \\ w \end{pmatrix} | w \in \mathbb{R} \right\}$ and $F_1 = \left\{ \begin{pmatrix} w \\ 0 \end{pmatrix} | w \in \mathbb{R} \right\}$, respectively. So, for (C, A) to be observable, choose a C which does not belong to both F_1 and F_2. In particular, for $C = \begin{pmatrix} 1 & 2 \end{pmatrix}$, (C, A) is observable, whereas for $C = \begin{pmatrix} 1 & 0 \end{pmatrix}$, (C, A) is unobservable.

4.3 Controllability of nonlinear systems

In this section, we discuss the controllability of the nonlinear system (5). We assume that the following conditions are satisfied.

a. The linear part of system (5) is controllable and $\|\Phi(t,s)\| \leq m$, for all $t, s \in [t_0, t_1]$.

b. Let $b = \sup_{t_0 \leq t \leq t_1} \|B(t)\| \leq \infty$

c. The function f satisfies Caratheodery conditions, that is, $f(t,x)$ is measurable with respect to t for all $x \in \mathbb{R}^n$ and continuous with respect to x for almost all $t \in [t_0, t_1]$. Further, f is Lipshitz continuous, that is, there exists a constant $\alpha \geq 0$ such that $\|f(t,x) - f(t,y)\| \leq \alpha \|x - y\|$ for all $x, y \in \mathbb{R}^n$.

We can reduce the controllability problem to a solvability problem. For $x \in C([t_0, t_1]; \mathbb{R}^n)$, let us define

$$x_1 - \Phi(t_1, t_0)x_0 - \int_{t_0}^{t_1} \Phi(t_1, s) f(s, x(s)) ds = \sum_{i=1}^{n} c_{x_i} v_i \tag{28}$$

where $\{v_n\}$ is the orthonormal basis of $\mathbb{R}^n$ generated by the eigenvectors corresponding to the eigenvalues $\{\lambda_i\}$ of $\mathcal{W}(t_0, t_1)$. Here c'_{x_i}s are coordinates of the vector $x_1 - \Phi(t_1, t_0)x_0 - \int_{t_0}^{t_1} \Phi(t_1, t) f(t, x(t)) dt$ with respect to the orthonormal basis $\{v_n\}$. Now, the control defined by

$$u(t) = B^T(t) \Phi^T(t_1, t) \sum_i \frac{c_{x_i} v_i}{\lambda_i} \tag{29}$$

steers the system (5) from x_0 to x_1 during $[t_0, t_1]$, provided x in (26) satisfies

$$x(t) = \Phi(t, t_0)x_0 + \int_{t_0}^{t} \Phi(t, s)B(s)B^T(s)\Phi^T(t_1, s)\sum_i \frac{c_{x_i}v_i}{\lambda_i} ds + \int_{t_0}^{t} \Phi(t, s)f(s, x(s))ds \tag{30}$$

We apply Banach fixed point theorem for establishing the solvability of the Eq. (30). We define a mapping $F : C([t_0, t_1]; \mathbb{R}^n) \to C([t_0, t_1]; \mathbb{R}^n)$ by

$$(Fx)(t) = \Phi(t, t_0)x_0 + \int_{t_0}^{t} \Phi(t, s)B(s)B^T(s)\Phi^T(t_1, s)\sum_i \frac{c_{x_i}v_i}{\lambda_i} ds + \int_{t_0}^{t} \Phi(t, s)f(s, x(s))ds \tag{31}$$

The solvability of Eq. (30) follows if we prove that F has a fixed point. We first prove the following lemmas.

Theorem 9 [8]: Under the assumptions (a) and (c), we have the following inequality:

$$\left\| \sum_{i=1}^{n} \frac{c_{x_i}v_i}{\lambda_i} - \sum_{i=1}^{n} \frac{c_{y_i}v_i}{\lambda_i} \right\| \le \frac{1}{|\lambda|} m\alpha \int_{t_0}^{t_1} \|y(s) - x(s)\| ds$$

where $|\lambda| = min\{|\lambda_1|, |\lambda_2|, \ldots, |\lambda_n|\} \neq 0, x, y \in C([t_0, t_1]; \mathbb{R}^n)$ and $\{v_n\}$ is the orthonormal basis of $\mathbb{R}^n$ generated by the eigenvectors corresponding to the eigenvalues $\{\lambda_i\}$ of $\mathcal{W}(t_0, t_1)$. Hence, c'_{x_i}s are coordinates of the vector $x_1 - \Phi(t_1, t_0)x_0 - \int_{t_0}^{t_1} \Phi(t_1, t)f(t, x(t))dt$ with respect to the orthonormal basis $\{v_n\}$ and c'_{y_i}s are coordinates of the vector $x_1 - \Phi(t_1, t_0)x_0 - \int_{t_0}^{t_1} \Phi(t_1, t)f(t, y(t))dt$ with respect to the orthonormal basis $\{v_n\}$.

Proof: We have

$$\begin{aligned}
\left\| \sum_{i=1}^{n} \frac{c_{x_i}v_i}{\lambda_i} - \sum_{i=1}^{n} \frac{c_{y_i}v_i}{\lambda_i} \right\| &\le \left\| \sum_{i=1}^{n} \frac{c_{x_i}v_i - c_{y_i}v_i}{\lambda_i} \right\| \\
&\le \frac{1}{|\lambda|} \left\| \sum_i \left(c_{x_i}v_i - c_{y_i}v_i\right) \right\| \\
&= \frac{1}{|\lambda|} \left\| \int_{t_0}^{t_1} \Phi(t_1, s)f(s, y(s))ds - \int_{t_{01}}^{t} \Phi(t_1, s)f(s, x(s))ds \right\| \\
&\le \frac{1}{|\lambda|} \int_{t_0}^{t_1} \|\Phi(t_1, s)\| \left\| f(s, y(s)) - f(s, x(s)) \right\| ds \le \frac{1}{|\lambda|} m\alpha \int_{t_0}^{t_1} \|y(s) - x(s)\| ds
\end{aligned}$$

Theorem 10: Under the assumptions (a), (b), and (c) and

$$\alpha m(t_1 - t_0)\left(\frac{m^2b^2}{|\lambda|}(t_1 - t_0) + 1\right) < 1$$

the operator F is a contraction.

Proof: Let x and y be solutions of system (5). Then,

$$
\begin{aligned}
\|Fx-Fy\| = & \sup_{t\in[t_0,t_1]}\left\|\int_{t_0}^{t}\Phi(t,s)B(s)B^T(s)\Phi^T(s)\left(\sum_i\frac{c_{x_i}v_i}{\lambda_i}-\sum_i\frac{c_{y_i}v_i}{\lambda_i}\right)ds+\int_{t_0}^{t}\Phi(t,s)(f(s,x(s))-f(s,y(s)))\right\|\\
&\le \sup_{t\in[t_0,t_1]}\left\|\int_{t_0}^{t}\Phi(t,s)B(s)B^T(s)\Phi^T(s)\left(\sum_i\frac{(c_{x_i}-c_{y_i})v_i}{\lambda_i}\right)ds\right\|\\
&+\sup_{t\in[t_0,t_1]}\left\|\int_{t_0}^{t}\Phi(t,s)(f(s,x(s))-f(s,y(s)))\right\|\\
&\le \sup_{t\in[t_0,t_1]}\int_{t_0}^{t}\|\Phi(t,s)\|\,\|B(s)\|\|B^T(s)\|\|\Phi^T(s)\|\left\|\sum_i\frac{(c_{x_i}-c_{y_i})v_i}{\lambda_i}\right\|ds\\
&+\sup_{t\in[t_0,t_1]}\int_{t_0}^{t}\|\Phi(t,s)\|\|f(s,x(s))-f(s,y(s))\|\\
&\le \sup_{t\in[t_0,t_1]} m^2b^2\int_{t_0}^{t}\left|\sum_i\frac{(c_{x_i}-c_{y_i})v_i}{\lambda_i}\right|, m\sup_{t\in[t_0,t_1]}\int_{t_0}^{t}\alpha\|x(s)-y(s)\|ds\\
&\le \sup_{t\in[t_0,t_1]} m^2b^2\int_{t_0}^{t}\frac{1}{|\lambda|}m\alpha\int_{t_01}^{t}\|y(\tau)-x(\tau)\|d\tau ds+m\sup_{t_1]}\int_{t_0}^{t}\alpha\|x(s)-y(s)\|ds\\
&\le m^2b^2\frac{1}{|\lambda|}m\alpha(t_1-t_0)^2\|y-x\|+m\alpha(t_1-t_0)\|y-x\|\\
&\le m\alpha(t_1-t_0)\left(\frac{m^2b^2}{|\lambda|}(t_1-t_0)+1\right)\|y-x\|
\end{aligned}
$$

Since $\alpha \le \frac{1}{m(t_1-t_0)\left(\frac{m^2b^2}{|\lambda|}(t_1-t_0)+1\right)}$, F is a contraction.

Theorem 11 [8]: The system (5) is controllable if $A(t)$ and $B(t)$ satisfy assumptions (a), (b) and the function f satisfies the assumption (c) and is Lipchitz continuous with Lipshitz constant

$$\alpha \le \frac{1}{m(t_1-t_0)\left(\frac{m^2b^2}{|\lambda|}(t_1-t_0)+1\right)}$$

An algorithm for the computation of steering control and controlled trajectory is given by:

$$u^n(t)=B^T(t)\Phi^T(t_1,t)\sum_i\frac{c^n_{x_i}v_i}{\lambda_i} \tag{32}$$

$$x^{n+1}(t)=\Phi(t,t_0)x_0+\int_{t_0}^{t}\Phi(t,s)B(s)u^n(s)ds+\int_{t_0}^{t}\Phi(t,s)f(s,x^n(s))ds \tag{33}$$

where $x^0(t)=x_0,\ \ n=1,2,3,\ \dots$.

Proof: By Lemma 11, the operator F is a contraction. Hence by Banach contraction principle F has a unique fixed point. Thus, the Eq. (28) is solvable. And thus the nonlinear system is controllable. The computational algorithm follows directly from Banach contraction principle.

5. Conclusions

Eigenvalues and eigenvectors of the matrices associated with a control system play a significant role in identifying control theoretic properties. The controllability and observability of a system can be determined with the aid of eigenvectors and the idea of orthogonality, this is the content of the famous PBH theorem. Apart from qualitative analysis of the system, eigenvalues and eigenvectors can be employed in computing the steering control. The implications of spectral studies in system design, as dealt with in the examples in the text, are noteworthy. The plant, input, and observation matrices can be chosen so as to make the system controllable and observable. In this chapter, we have dealt with systems in the state-space form, whereas there are more applications of eigenvalues and eigenvectors in the transfer-function approach.

A. Appendix

The Kalman's rank conditions for controllability and observability of LTI systems were proposed by the Hungarian-American electrical engineer, mathematician, and inventor *Rudolf E. Kálman (1930–2016)*.

Theorem 12: If the system (4) is LTI, then it is controllable if and only if the controllability matrix

$$Q(A,B) = \left[B|AB|\cdots|A^{n-1}B\right]$$

is of full rank. That is, $Rank\,(Q(A,B)) = n$.

Proof: Suppose that the system (4) is controllable. That is, $Rangespace(\mathcal{C}) = \mathbb{R}^n$. As $Q(A, B)$ can be considered as a bounded linear operator from $\mathbb{R}^{mn} \to \mathbb{R}^n$, to show that Q is of full rank it is enough to prove that $Rangespace\,[Q(A,B)] = \mathbb{R}^n$. Clearly, $Rangespace\,[Q(A,B)] \subset \mathbb{R}^n$. Now, let $v \in \mathbb{R}^n$. By Theorem 1, there exists $u \in \mathscr{L}^2([t_0,t_1];\mathbb{R}^m)$ such that $\mathcal{C}u = v$

$$\mathcal{C}u = v \Rightarrow \int_{t_0}^{t_1} \Phi(t_1,\tau)Bu(\tau)d\tau = v \Rightarrow \int_{t_0}^{t_1} e^{A(t_1-t)}Bu(\tau)d\tau = v$$

Expanding $e^{A(t_1-t)}$ and by using Cayley-Hamilton theorem, we have

$$\int_{t_0}^{t_1} \left[P_0(\tau)I + P_1(\tau)A + \ \dots \ + P_{n-1}(\tau)A^{n-1}\right]Bu(\tau)d\tau = v$$

where each $P_i(\tau)$ is a polynomial function of τ that appears during the expansion of $e^{A(t_1-t)}$. This implies that $v \in Rangespace\,[Q(A,B)]$. Therefore, $\mathbb{R}^n \subset Rangespace[Q(A,B)]$ and hence $Rank[Q(A,B)] = n$.

Conversely suppose that the system (4) is not controllable. Then by Theorem 1 $\mathcal{W}(t_0,t_1)$ is not invertible and hence there exists $v \neq 0 \in \mathbb{R}^n$ such that $\mathcal{W}(t_0,t_1)v = 0$. This implies that $v^T\mathcal{W}(t_0,t_1)v = 0$. Therefore,

$$\langle \mathcal{W}v, v\rangle = \left\langle \int_{t_0}^{t_1} e^{A(t_1-t)}BB^Tu(\tau)e^{A^T(t_1-t)}v d\tau, v \right\rangle = 0$$

This implies that

$$\int_{t_0}^{t_1} v^T e^{A(t_1-t)} B B^T e^{A^T(t_1-t)} v = \int_{t_0}^{t_1} \|B^T e^{A^T(t_1-t)} v\|^2 = 0$$

As $B^T e^{A^T(t_1-t)} v$ is a continuous function on $[t_0, t_1]$, this implies

$$B^T e^{A^T(t_1-t)} v = 0, \forall\ t \in [t_0, t_1] \Rightarrow v^T e^{A(t_1-t)} B = 0, \forall\ t \in [t_0, t_1]$$

In particular, for $t = t_1$, $v^T B = 0$. Further, differentiating $v^T e^{A(t_1-t)} B$ w.r.t. t and evaluating at $t = t_1$, we get $v^T AB = 0$. Successively differentiating and evaluating at $t = t_1$, we get

$$v^T B = v^T AB = \cdots = v^T A^{n-1} B = 0$$

That is, $v \perp Rangespace\left([B|AB|\cdots|A^{n-1}B]\right)$. This implies that $Rank\,[B|AB|\cdots|A^{n-1}B] < n$. Hence, the result follows by contraposition.

Theorem 13: If the system (20) is LTI, then it is observable if and only if the observability matrix

$$\mathcal{O}(C, A) = rank \begin{bmatrix} C \\ CA \\ \vdots \\ CA^{n-1} \end{bmatrix}$$

has full column rank. That is, $rank[\mathcal{O}(C, A)] = n$.

Proof: Proof follows from duality and Kalman's rank condition for controllability.

B. Appendix

B.1 MATLAB code for Example 3

```
syms t;
A=[0 1;-25 -10]; % state matrix
B=[5;1]; % control matrix
x0=[0;0]; % initial state
x1=[1;1]; % final state
Phi=expm(A*t); % transition matrix
Wc=gram(ss(A,B,[],[]),'c'); % Gramian
disp(Wc);
[Vw,Dw]=eig(Wc); % Eigenvalues and Eigenvectors of Gramian
c=inv(Vw)*x1;
s1=(c(1)*Vw(:,1))/Dw(1,1);
s2=(c(2)*Vw(:,2))/Dw(2,2);
u=transpose(B)*transpose(expm(A*(1-t)))*(s1+s2) % Control input computation
syms y
aa=int(expm(A*(1-t))*B*u);
```

```
y=expm(A*t)*x0+aa % Trajectory computation
tiledlayout(2,2)
nexttile
fplot(u,[0,1],'g')
nexttile
fplot(y(1),[0,1],'r')
hold on
fplot(y(2),[0,1],'b')
legend('x_1(t))','x_2(t))','Location','northwest')
```

Author details

Raju K. George* and Abhijith Ajayakumar
Department of Mathematics, Indian Institute of Space Science and Technology, Valiyamala, Trivandrum, India

*Address all correspondence to: rkg.iist@gmail.com

References

[1] Chalishajar DN, George RK, Nandakumaran AK, Acharya FS. Trajectory controllability of nonlinear integro-differential system. Journal of the Franklin Institute. 2010;**347**(7): 1065-1075

[2] George RK, Ajayakumar A. A Course in Linear Algebra. Springer; 2024

[3] Saikia PK. Linear Algebra, 2e. Pearson Education India; 2014

[4] Brockett RW. Finite Dimensional Linear Systems. Society for Industrial and Applied Mathematics; 2015

[5] Klamka J. Controllability and Minimum Energy Control. Springer International Publishing; 2019

[6] Sontag ED. Mathematical Control Theory: Deterministic Finite Dimensional Systems. Vol. 6. Springer Science & Business Media; 2013

[7] Terrell WJ. Stability and Stabilization: An Introduction. Princeton University Press; 2009

[8] Sharma JP, George RK. In: Balasubramaniam et al., editors. Controllability and Steering Control by Spectral Method, Mathematics, Computing and Modelling. Chennai: Allied Publishers; 2007. pp. 11-20

Chapter 3

Solution for Matrix Exponentiation Using Eigenvalues

Dawit Hiluf Hailu

Abstract

In this chapter, we introduce the application of Sylvester's formula for systems with degenerate eigenvalues to obtain their analytical solutions. To illustrate its utility, we include two additional methods for analytical solutions: the adiabatic approximation and the Magnus expansion. In quantum mechanics, the Schrödinger equation is a mathematical expression that describes the time evolution of a physical system where quantum effects, such as wave-particle duality, are significant. This equation serves as a framework for analyzing quantum mechanical systems. Similar to how Newton's laws govern the motion of classical objects, the Schrödinger equation governs the behavior of quantum entities. However, unlike classical motion, quantum mechanics deals with the probabilities of different trajectories. The ability to obtain analytical solutions to such equations provides an in-depth understanding of the dynamics and also allows for the identification of controlling parameters, as analytical functions can readily reveal these insights.

Keywords: matrix exponent, eigenvalues, eigenvectors, Sylvester formula, degenerate eigenvalues

1. Introduction

Generally speaking, physical systems are dynamic, meaning they evolve. The dynamical evolution of quantum mechanical systems is described by the Schrödinger equation, which characterizes their dynamic properties. Given an initial quantum state $|\psi(0)\rangle$ at time $t = 0$, a physical law is required to determine the state at any subsequent time t, that is, a quantum law of evolution. Since a state vector contains all known information about a system, we need a general physical law to describe how this information evolves in response to the specific physical conditions affecting the system. This exploration focuses on how the system responds to external perturbations. Although the details of this law may vary between systems, it can be expressed in a universally applicable form.

The dynamics of quantum mechanical systems without external perturbations can be effectively studied using the Schrödinger equation, assuming a constant Hamiltonian. Solutions to such systems are often, though not always, achievable through analytical approaches. However, this typically requires making assumptions or approximations about the system, simplifying its complexity to allow for analytical solutions. These simplifications help focus on the relevant aspects of the system, excluding extraneous information. Obtaining such solutions aids in identifying controlling parameters, which is valuable for experimental design and understanding the system under investigation. According to Bohr, such states are known as stationary states, referring to quantum states that do not change over time, though they still possess significant physical properties.

Our aim in this chapter is to demonstrate the use of Sylvester's formula for dynamical systems with degenerate eigenvalues. We will first introduce some approximations and assumptions that justify the use of analytical solutions. Among the various methods for obtaining analytical solutions, we have chosen Sylvester's formula because it is particularly suitable for systems with degenerate eigenvalues. Sylvester's formula requires only the knowledge of eigenvalues, making it an ideal candidate for our purposes.

Our approach to obtaining solutions *via* eigenvalues is grounded in quantum mechanics. In this context, a quantum state can be viewed as a vector in Hilbert space, with measurements on the state described as operations or time evolutions, such as those induced by interacting laser light. This can be represented as a matrix acting on a state vector, where the eigenvalues correspond to the energy of the state. This description applies to any observable or physically measurable quantity in quantum mechanics. It is the eigenvalues of these observables that correspond to the measured values.

Strictly speaking, observations in quantum physics do not typically yield the observables themselves as outcomes but rather the expectation values of these observables. This indicates that the connection between theory and experiment is through the expectation values of the observables. It is important to note that these expectation values are real numbers corresponding to the outcomes of measuring the observables. We work in the Heisenberg picture, where observables are time-dependent, specifying the changing expectation values of the observables. Mathematically, the expectation values are given by a fixed linear function, $\langle O \rangle$, mapping observables to real numbers.

To reiterate, in quantum mechanics, the Schrödinger equation is a mathematical formulation that describes the time evolution of a physical system where quantum effects are significant. The Schrödinger equation of motion is used to study quantum mechanical systems. This approach requires knowledge of the Hamiltonian and the initial state of the system at $t = t_0$ to determine the state of the quantum system at a later time t. The Hamiltonian, typically denoted by H or $\hat{H}$, corresponds to the total energy of the system. Its spectrum consists of the possible outcomes when measuring the system's total energy. Due to its close relation to the time evolution of a system, the Hamiltonian is fundamentally important in most formulations of quantum theory, generating the time evolution of quantum states.

Each measurable parameter in a physical system has an associated quantum mechanical operator, with the Hamiltonian being the operator associated with the system's energy. The Hamiltonian includes the operations associated with kinetic and potential energies. Applying the Hamiltonian to the wave function produces the Schrödinger equation. In the time-independent Schrödinger equation, this operation yields specific energy values known as energy eigenvalues. Beyond determining system energies, the Hamiltonian operator also generates the time evolution of the wave function.

In this chapter, we will explore the Hamiltonian of a system interacting with an electric field. We then use the Hamiltonian to describe the system's equation of motion. The evolution of the system can be described either through the probability amplitude and the time-dependent Schrödinger equation or using the density matrix formalism in Liouville's description. Using the equations for the density matrix elements, we will form a set of observables through a linear combination of the coherences and populations. We then obtain the equation of motion in a larger space and seek an analytical solution, assuming the perturbation fields have the same time dependence but potentially different strengths.

2. Analytical approximations

Once the equations of motion for quantum systems are established, it is evident that they can be solved numerically. However, our focus in this chapter is on exploring some of the analytical approaches to solving these equations. It is important to recognize that these analytical solutions are approximations and may not be as precise as those obtained through numerical methods. Nonetheless, they often provide valuable physical insights.

The systems of equations we aim to address in this chapter are coupled equations of motion, as fully derived in such as Refs. [1–3]. These equations take the form of the following equation of motion (see Eq. (1)) for the observable vector $\vec{S}$. The observable vector $\vec{S}$ consists of the expectation values of the observables, which are linear combinations of the coherences and populations. Since the observable vector exists in a larger space, it contains more elements. For a quantum system with N distinct states, the vector has $N^2 - 1$ elements if normalization is imposed. For example, in a two-level system, the observable vector has three elements and is commonly referred to as the Bloch vector [4, 5].

A key aspect of the derivation is that the Hamiltonian and the density matrix can be expressed as linear combinations of generators, thereby forming a closed Lie algebra. It is worth noting that the discussion in this chapter is equally applicable to time-dependent Schrödinger equations in Hilbert space, as well as other equations of motion with a similar form.

$$\frac{d}{dt}\vec{S} = g\vec{S} \tag{1}$$

In contrast to the Schrödinger equation, the vector $\vec{S}$ in Eq. (1) describes the state of the system, while the matrix g represents its Hamiltonian, with dimensions $(N^2 - 1) \times (N^2 - 1)$. As mentioned earlier, such equations can be solved numerically to obtain exact solutions. However, there are instances when approximate solutions are necessary. These approximations can provide valuable insights into the physical system despite the loss of exactness. We emphasize that by adjusting all or selected parameters, we can approach the exact solution more closely. Furthermore, analytical calculations can be useful, and sometimes preferable, for identifying the most essential or controlling parameters. Understanding these controlling parameters is crucial for proposing and designing experimental setups.

In this section, before we delve into using Sylvester's formula to obtain analytical solutions, we will first examine two other methods for obtaining analytical solutions:

the adiabatic approximation and the Magnus expansion. We will provide a brief overview of these two methods.

2.1 Adiabatic approximations

In the context of this discussion, adiabaticity or adiabatic approximation refers to the scenario where the perturbation on a physical system changes very slowly over time. This slow change allows the system to align with the same eigenstate before and after the interaction with the perturbation. Because the variation in perturbation is gradual, the system has ample time to adjust to its instantaneous eigenstate. Essentially, this means that once the system is prepared in an instantaneous eigenstate, it remains in that state as long as its eigenvalue is separated from the nearest states by a finite energy gap. Physically, this implies there is no transition between adiabatic states. This characteristic is reflected in the structure of the adiabatic Hamiltonian, where the matrix elements of the transformed Hamiltonian (the Hamiltonian in the adiabatic picture) are zero or negligibly small, except on its diagonal.

Mathematically, this can be achieved by considering a unitary transformation matrix U that diagonalizes the matrix g in Eq. (1). The adiabatic approximation is a method used when the Hamiltonian of the system changes slowly compared to the evolution of the system's state. This approach assumes that the system remains in its instantaneous eigenstate as the Hamiltonian varies. The key idea is that if the changes in the Hamiltonian are slow enough, the system can adjust its state adiabatically, meaning there is no transition between different eigenstates. This method is particularly useful for systems where the Hamiltonian varies due to external influences such as slowly changing fields.

The adiabatic approximation simplifies the problem by reducing the time-dependent Schrödinger equation to a set of simpler equations that are easier to solve. This method provides a good approximation when the conditions of slow variation are met and can give significant insights into the system's behavior under slowly changing conditions.

Following transformation rules, let us now multiply both sides of Eq. (1) by a unitary transformation matrix U^{-1} whose time derivative is zero.

$$U^{-1}\frac{d}{dt}\vec{S} = U^{-1}g\vec{S} \tag{2}$$

$$\frac{d}{dt}\left(U^{-1}\vec{S}\right) = U^{-1}g\vec{S} \tag{3}$$

By multiplying the right-hand side (RHS) of Eq. (3) by the identity matrix $I = UU^{-1}$, the matrix g can be readily diagonalized. This transformation results in a diagonal matrix $\Lambda = U^{-1}gU$, whose elements are the eigenvalues of g.

$$\frac{d}{dt}\left(U^{-1}\vec{S}\right) = \Lambda\left(U^{-1}\vec{S}\right) \tag{4}$$

For convenience, let us denote the transformed observable vector as $\vec{S}' = U^{-1}\vec{S}$. Therefore, the equation of motion for this new observable vector is

$$\frac{d}{dt}\vec{S}' = \Lambda \vec{S}' \tag{5}$$

In this approximation, the transformed observable vector $\vec{S}'$ evolves according to a simplified set of equations where the Hamiltonian is diagonal. This allows us to express the solution in terms of the eigenvalues of the Hamiltonian matrix. Recalling that we are using the adiabatic approximation, the solution takes the form

$$\vec{S}'(t) = e^{\Lambda t}\vec{S}'(0) \tag{6}$$

Since we are seeking the solution for Eq. (1), we can obtain it from the solution in Eq. (6) by multiplying the final equation by U, thereby transforming back to the original observable vector $\vec{S}(t) = U\vec{S}'(t)$

$$U\vec{S}'(t) = Ue^{\Lambda t}\vec{S}'(0) \tag{7}$$

$$\vec{S}(t) = Ue^{\Lambda t}U^{-1}\vec{S}(0) \tag{8}$$

By inserting the identity matrix on the RHS of Eq. (7), we arrive at Eq. (8). This solution can be expressed component-wise as follows:

$$S_k(t) = \sum_j U_{kj} e^{\lambda_j t} U^{-1}_{jk} \vec{S}_j(0) \tag{9}$$

where λ_j is the eigenvalue.

Example - i: Let us illustrate the adiabatic approximation method by applying it to a two-level system and deriving the solution for the coherence vector, or Bloch vector. Consider a two-level system with ground state $|0\rangle$ and excited state $|1\rangle$, coupled by a coherent laser light of Rabi frequency $\hbar\Omega$ and detuning $\Delta = \hbar(\omega_1 - \omega_0)$. Assume the system is initially prepared in the ground state.

Solution: To begin, we construct the coherence vector using generators that form a closed Lie algebra, as discussed in Refs. [4, 5] and related references. For clarity and future reference, we restate here the relationship between the generators and the components of the coherence vectors as

$$\begin{aligned} u_{01} &= |0\rangle\langle 1| + |0\rangle\langle 1|, \\ v_{01} &= -i(|0\rangle\langle 1| - |0\rangle\langle 1|), \\ w_1 &= |0\rangle\langle 0| - |1\rangle\langle 1| \end{aligned} \tag{10}$$

The equation of motion for a two-level system in terms of $SU(2)$ symmetry takes the following form:

$$\begin{aligned} \frac{d}{dt}\begin{pmatrix} S_1 \\ S_2 \\ S_3 \end{pmatrix} &= \begin{pmatrix} 0 & \Delta & 0 \\ -\Delta & 0 & -\Omega \\ 0 & \Omega & 0 \end{pmatrix}\begin{pmatrix} S_1 \\ S_2 \\ S_3 \end{pmatrix} \\ \frac{d}{dt}\vec{S} &= g\vec{S} \end{aligned} \tag{11}$$

One can readily obtain the eigenvalues and eigenvectors of g as follows

$$\Lambda = \begin{pmatrix} 0 & 0 & 0 \\ 0 & -\sqrt{-\Delta^2-\Omega^2} & 0 \\ 0 & 0 & \sqrt{-\Delta^2-\Omega^2} \end{pmatrix}$$
$$U = \begin{pmatrix} -\frac{\Omega}{\Delta} & \frac{\Delta}{\Omega} & \frac{\Delta}{\Omega} \\ 0 & -\frac{\sqrt{-\Delta^2-\Omega^2}}{\Omega} & \frac{\sqrt{-\Delta^2-\Omega^2}}{\Omega} \\ 0 & \Omega & 0 \end{pmatrix} \tag{12}$$

Therefore, by substituting the values and noting that $\vec{S}(0) = (0,0,1)^T$, reflecting the initial preparation of the system in the ground state, where $\vec{S} = (u_{01}, v_{01}, w_1)^T$, and performing some algebraic manipulation, we arrive at the following result:

$$\vec{S}(t) = Ue^{\Lambda t}U^{-1}\vec{S}(0)$$
$$\vec{S}(t) = -\begin{pmatrix} \left(-1+\frac{1}{2}\left(e^{\lambda_1 t}+e^{\lambda_2 t}\right)\right)\frac{\Delta\Omega}{\Delta^2+\Omega^2} \\ \frac{1}{2}\left(-e^{\lambda_1 t}+e^{\lambda_2 t}\right)\frac{\Omega\sqrt{-\Delta^2-\Omega^2}}{\Delta^2+\Omega^2} \\ \left(\Delta^2+\frac{\Omega^2}{2}\left(e^{\lambda_1 t}+e^{\lambda_2 t}\right)\right)\frac{1}{\Delta^2+\Omega^2} \end{pmatrix} \tag{13}$$

If we introduce $\Omega_0 = \sqrt{\Delta^2+\Omega^2}$, which gives $\lambda_1 = -\sqrt{-\Delta^2-\Omega^2} = -i\Omega_0$, similarly, we obtain $\lambda_2 = \sqrt{-\Delta^2-\Omega^2} = i\Omega_0$, see **Figure 1**. With these definitions, the solutions can be rewritten as

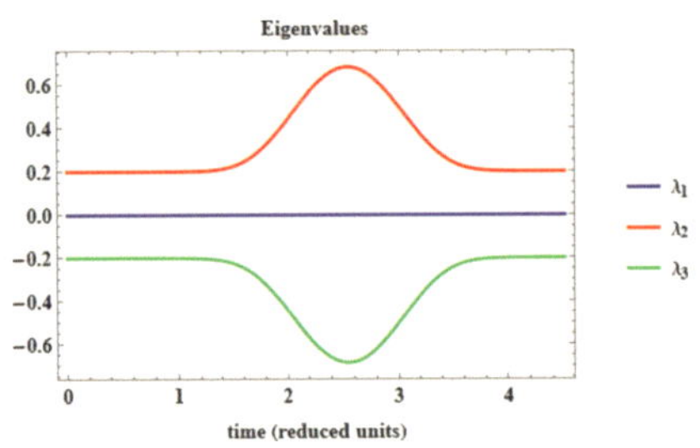

Figure 1.
Eigenvalues for a two-level system with coupling laser in the reduced unit.

$$\vec{S}(t) = \begin{pmatrix} \dfrac{\Delta\Omega}{\Omega_0^2}(1 - \cos\Omega_0 t) \\ \dfrac{\Omega}{\Omega_0}\sin\Omega_0 t \\ -\dfrac{\Delta^2}{\Omega_0^2} - \dfrac{\Omega^2}{\Omega_0^2}\cos\Omega_0 t \end{pmatrix} \tag{14}$$

To leverage the advantages of having an analytical solution, let us pause and examine what insights we can gain from the solution obtained in Eq. (14). Firstly, based on the definition used to construct the coherence vector, as seen in Eq. (10), it is important to note that the solution describes the evolution of the vector at time t after the system has been perturbed by a laser field. The elements of the coherence vector consist of the real and imaginary parts of the coherences (i.e., ρ_{01} of the density matrix) and the population differences between the two levels (i.e., ρ_{aa} for $a = 0, 1$).

Furthermore, we use this solution as a starting point for our objectives. For example, considering the well-known example of complete population transfer in a two-level system, our goal would be to achieve $\vec{S}(t) = (0, 0, -1)^T$, see **Figure 2**. Referring to Eq. (14), this requires that

$$\begin{aligned} \frac{\Delta\Omega}{\Omega_0^2}(1 - \cos\Omega_0 t) &= 0 \\ \frac{\Omega}{\Omega_0}\sin\Omega_0 t &= 0 \\ -\frac{\Delta^2}{\Omega_0^2} - \frac{\Omega^2}{\Omega_0^2}\cos\Omega_0 t &= -1 \end{aligned} \tag{15}$$

It is evident that the solution depends on Δ and Ω, as well as their combinations. From these equations, we observe that the desired condition is met when $\Omega_0 t = \pm\pi$ for a negligibly small detuning $\Delta \to 0$. This condition implies $\Omega t = \pm i\pi$. Specifically, on resonance where $\Delta = 0$, the solution takes the form

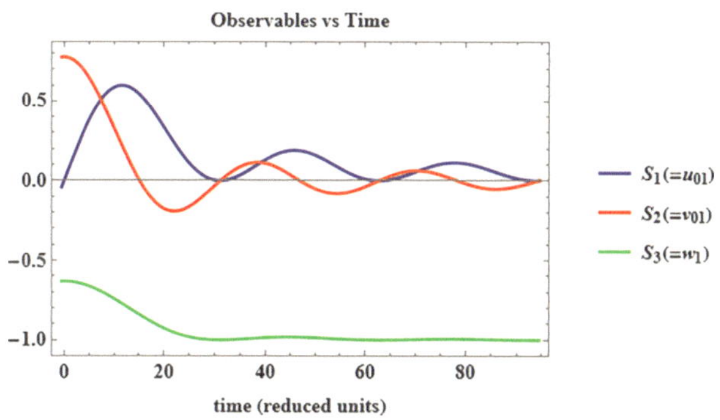

Figure 2.
The result shows a complete population inversion as outlined in the text. Solution is obtained using Eq. (49).

$$\vec{S}(t) = -\begin{pmatrix} 0 \\ \frac{i}{2}\left(e^{-i\Omega t} - e^{i\Omega t}\right) \\ -\frac{1}{2}\left(e^{-i\Omega t} + e^{i\Omega t}\right) \end{pmatrix} \tag{16}$$

Let us now make a dimensional analysis and work with dimensionless quantities for simplicity. To this end, it is helpful to work with reduced units. Reduced time is a concept used in various fields of science and engineering to simplify the analysis of dynamic systems. Scaling the time variable often helps to make equations dimensionless or to compare systems with different time scales more easily.

In the context of our scenario involving laser pulses and quantum states, reduced time is used to normalize the time variable based on characteristic timescales of the system, such as the duration of the laser pulse or the natural oscillation period of the system.

For example, if t is the actual time and σ is a characteristic timescale (like the duration of a laser pulse), the reduced time τ can be defined as:

$$\tau = \frac{t}{\sigma} \tag{17}$$

This way, τ becomes a dimensionless variable that simplifies the mathematical treatment of the problem. When dealing with differential equations, this normalization often helps to reduce the number of parameters and reveals the underlying structure of the equations more clearly.

In our context of population transfer between quantum states, reduced time could help in comparing how different laser pulses (with different durations or intensities) affect the population dynamics in a standardized way. To this end, we keep on using the reduced time scale throughout.

Utilizing and $\lambda_2 = \sqrt{-\Delta^2 - \Omega^2} = i\Omega_0$, the expression becomes

$$\vec{S}(t) = \begin{pmatrix} 0 \\ \sin \Omega_0 t \\ -\cos \Omega_0 t \end{pmatrix} \tag{18}$$

If the goal, on the other hand, is to induce coherences between the ground and excited states—achieving equal population distribution between the two states—the desired outcomes would be either $\vec{S}(t) = \left(\frac{1}{2},0,0\right)^T$ or $\vec{S}(t) = \left(0, \frac{1}{2}, 0\right)^T$, depending on the nonzero real or imaginary part of the off-diagonal element in the density matrix. The solution from Eq. (14) imposes several conditions to achieve this, notably $\Omega t = \pm\frac{\pi}{2}$ and $\Delta t = \pm\frac{\pi}{2}$.

Following similar reasoning as above, one can derive various other solutions depending on the desired population distributions, such as transferring a third or a quarter of the population to the excited state or vice versa. These solutions can guide experimentalists in designing and conducting experiments that meet specific conditions. The condition $|\Omega t|$, known as the pulse area, indicates the strength of the laser pulse: a pulse area of π corresponds to a strong laser field, while $\frac{\pi}{2}$ indicates a weaker applied laser field.

To summarize, the analytical approach provides valuable insights into the physics of the system under study. It allows us to predict and understand how different experimental parameters, such as pulse area, affect the quantum dynamics, thereby guiding experimental design and interpretation.

2.2 Sylvester's formula

The approach outlined in subsection (2.1) assumes that the Hamiltonian of the system commutes with itself at different times. However, in the case of ordinary differential equations (ODEs) with time-varying coefficients, such as the one under consideration here,

$$\frac{d}{dt}\vec{S}(t) = g(t)\vec{S}(t), \quad \vec{S}(0) = S_0 \tag{19}$$

with $\vec{S}(t) = (S_1(t) \dots, S_n(t))$ and $g(t)$ is an $n \times n$ matrix, (the general solution) is given by

$$\vec{S}(t) = \mathcal{T}\left\{e^{\int_0^t g(t')dt'}\right\}\vec{S}(0), \tag{20}$$

where $\mathcal{T}$ denotes time-ordering,

$$\begin{aligned} \mathcal{T}\left\{e^{\int_0^t g(t')dt'}\right\} &\equiv \sum_{n=0}^{\infty}\frac{1}{n!}\int_0^t \dots \int_0^t \mathcal{T}\{g(t_1')\dots g(t_n')\} \\ &= \sum_{n=0}^{\infty}\int_0^t dt' \dots \int_0^{t_{n-1}'} dt_n' \, g(t_1')\dots g(t_n') \end{aligned} \tag{21}$$

Assuming the matrices commute at different times, meaning the commutator $[g(t_1), g(t_2)] = 0$ for all t_1, t_2, the time-ordered expression takes the form $e^{\int_0^t g(t')dt'}$. The evaluation of such exponential is extensively studied [6–10], but is beyond the scope of the present paper.

For example, the dynamics of the two-level system investigated in this chapter are approached using a Magnus series [4–6]. However, one can employ various proposed methods from the references to approximate or exactly solve these equations. This section uses Sylvester's formula to derive analytical solutions for the coupled differential equations.

2.2.1 Sylvester's formula for distinct eigenvalues

Another method to solve Eq. (1) involves Sylvester's formula, a widely recognized tool for solving exponential equations. Specifically, given an $N \times N$ coefficient matrix $g(t)$, our objective is to solve the initial value problem related to the linear ordinary differential equation governing the coherence vectors' equation of motion (refer to (1)). For clarity and convenience, we rewrite this equation along with its initial condition as

$$\frac{d}{dt}\vec{S}(t) = g(t)\vec{S}(t), \quad \vec{S}(0) = S_0 \tag{22}$$

The solution can be expressed as

$$\vec{S}(t) = e^{\int_{t_0}^{t} g(t_1)dt_1}\vec{S}(0) \tag{23}$$

$$\vec{S}(t) = e^{G(t_1)}\vec{S}(0) \tag{24}$$

Let $G(t_1) = \int_0^t g(t_1')\,dt_1'$. We assume all entries g_{ij} of the matrix g are integrable functions. Consequently, we define the integral of the matrix as the matrix whose entries are integrals of the corresponding elements of g, expressed mathematically as $\int g(t)\,dt := \left(\int g_{ij}(t)\,dt\right)$. Sylvester's formula provides a method for computing the exponential of a matrix using only its eigenvalues [11, 12]. Therefore, for an $N \times N$ matrix, the solution to its exponential can be obtained by finding its eigenvalues and applying Sylvester's formula. Using this approach, we can express the exponent in Eq. (24) as

$$e^{G(t_1)} = \sum_{j=1}^{N^2-1} e^{\gamma_j} \prod_{j \neq k=1}^{N^2-1} \frac{G(t_1) - \gamma_j I}{\gamma_k - \gamma_j} \tag{25}$$

Let γ_j denote the eigenvalues of $G(t_1)$, I represent the identity matrix, and e^{γ_j} signify exponent of the eigenvalues.

It is important to note that Sylvester's formula can be applied in two distinct ways. The first approach involves directly substituting the matrix G and its eigenvalues into the formula, as indicated in Eq. (25). Here, the computation relies solely on knowing the eigenvalues of the matrix.

The second approach, which also requires knowledge of eigenvectors and eigenvalues, utilizes spectral decomposition. Below, we will illustrate both methods and derive the solution for a two-level system whose equation of motion corresponds to the Bloch equation. It is essential to emphasize that, for practical purposes and simplicity, we assume the matrix elements (specifically those of the Hamiltonian) to be integrable either analytically or numerically.

With Sylvester's Formula: Let us apply Sylvester's formula to derive the analytical solution of the coherence vector, or Bloch vector, for a two-level system. Similar to before, we assume the system is initially prepared in the ground state. The equation of motion for a two-level system in terms of $SU(2)$ can be expressed as

$$\frac{d}{dt}\begin{pmatrix} S_1 \\ S_2 \\ S_3 \end{pmatrix} = \begin{pmatrix} 0 & \Delta & 0 \\ -\Delta & 0 & -\Omega \\ 0 & \Omega & 0 \end{pmatrix}\begin{pmatrix} S_1 \\ S_2 \\ S_3 \end{pmatrix} \tag{26}$$

its solution, assuming the matrix g is commutable with itself at different times, can be written as

$$\vec{S}(t) = e^{G(t_1)}\vec{S}(0) \tag{27}$$

where $G(t_1) = \int_0^t g(t_1)dt_1$, and it explicitly is expressible as

$$\begin{aligned} G &= \int_0^t g(t_1)dt_1 = \int_0^t dt_1 \begin{pmatrix} 0 & \Delta & 0 \\ -\Delta & 0 & -\Omega \\ 0 & \Omega & 0 \end{pmatrix} \\ &= \begin{pmatrix} 0 & \int_0^t \Delta dt_1 & 0 \\ -\int_0^t \Delta dt_1 & 0 & -\int_0^t \Omega dt_1 \\ 0 & \int_0^t \Omega dt_1 & 0 \end{pmatrix} \\ &= \begin{pmatrix} 0 & \Delta' & 0 \\ -\Delta' & 0 & -\Omega' \\ 0 & \Omega' & 0 \end{pmatrix} \end{aligned} \tag{28}$$

where Δ' and Ω' are the integrated values of Δ and Ω, respectively.

The next step is to determine the eigenvalues of g. Since our matrix is low-dimensional, we can find the eigenvalues by solving the characteristic equation $\det(g - \lambda I) = 0$. This yields the eigenvalues $\left\{0, -\sqrt{-\Delta'^2 - \Omega'^2}, \sqrt{-\Delta'^2 - \Omega'^2}\right\}$. By substituting the initial values of the coherence vector $\vec{S}(0) = (0, 0, -1)^T$, which follows from the definition of the vector $\vec{S} = (u_{01}, v_{01}, w_1)^T$, we find that the solution can be expressed as

$$\begin{aligned} S(t) &= e^{G(t_1)}\vec{S}(0) = \left[\sum_{j=1}^{3} e^{\gamma_j} \prod_{j \neq k=1}^{3} \frac{G(t_1) - \gamma_j I}{\gamma_k - \gamma_j}\right] \cdot \vec{S}(0) \\ &= \left[e^{\gamma_1}\frac{(G - \gamma_2 I_{3\times3})(G - \gamma_3 I_{3\times3})}{(\gamma_1 - \gamma_2)(\gamma_1 - \gamma_3)} + e^{\gamma_2}\frac{(G - \gamma_1 I_{3\times3})(G - \gamma_3 I_{3\times3})}{(\gamma_2 - \gamma_1)(\gamma_2 - \gamma_3)} + e^{\gamma_3}\frac{(G - \gamma_1 I_{3\times3})(G - \gamma_2 I_{3\times3})}{(\gamma_3 - \gamma_1)(\gamma_3 - \gamma_2)}\right] \\ &\quad \cdot \vec{S}(0) \end{aligned} \tag{29}$$

Introducing $\zeta = \sqrt{\Delta'^2 + \Omega'^2}$ and recalling the eigenvalues to be $\gamma_1 = 0$, $\gamma_2 = -i\zeta$, and $\gamma_3 = i\zeta$, we substitute these values into Eq. (29). After performing some algebra, the solution takes the following form:

$$\vec{S}(t) = \begin{pmatrix} \frac{\Delta'\Omega'}{\zeta^2}(1 - \cos\zeta) \\ \frac{\Omega'}{\zeta}\sin\zeta \\ -\frac{\Delta'^2}{\zeta^2} - \frac{\Omega'^2}{\zeta^2}\cos\zeta \end{pmatrix} \tag{30}$$

Examining the solution in Eq. (30) reveals that it has a similar form to the one obtained using the adiabatic approximation, as seen in Eq. (14).

With Spectral decomposition: We use the spectral decomposition method here, which involves factorizing a matrix into its canonical form. This process represents the matrix in terms of its eigenvalues and eigenvectors. It is important to note that this method applies only to diagonalizable matrices. Recall that an matrix is diagonalizable if it has n independent eigenvectors.

$$e^{G(t_1)} = \sum Pe^{\gamma}P^{-1} \tag{31}$$

Since our matrix G (dropping the time argument for clarity) is diagonalizable, it follows that there exists an invertible matrix P such that $G = PDP^{-1}$, where D is a diagonal matrix containing the eigenvalues of G, and P is a matrix whose columns are the eigenvectors of G. In this case, $e^G = Pe^DP^{-1}$. We employed the Taylor series expansion of the exponential function along with the relation $\left(P^{-1}GP\right)^m = P^{-1}G^mP$. Thus, for eigenvectors η_j, the eigenvalue function is given by

$$G\eta_j = \gamma_j\eta_j \tag{32}$$

Also, using Frobenius covariant where $Q_j = \eta_j\eta_j^T$ we have:

$$e^G = \sum_j e^{\gamma_j}\eta_j\eta_j^T \tag{33}$$

The Frobenius covariants of our matrix are the projection matrices associated with its eigenvalues and eigenvectors. Using these covariants, the solution can be expressed as:

$$S(t) = \sum_j e^{\gamma_j}\eta_j\left(\eta_j^T S(0)\right) \tag{34}$$

the term $\eta_j^T S(0)$ represents a scalar product.

To reiterate, Sylvester's formula provides a method for computing the exponential of a matrix using only its eigenvalues. Therefore, for any $N \times N$ matrix, one can obtain the solution of its exponential by determining its eigenvalues and applying Sylvester's formula. We have expressed the Magnus solution for our coherence vector in terms of the eigenvectors as follows:

$$S(t) = \sum_j e^{\gamma_j}\eta_j\left(\eta_j^T S(0)\right) \tag{35}$$

If we define $Q_j = \eta_j \eta_j^T$ as the projection operator, where η_j is an eigenvector of the matrix, and Q_j satisfies the idempotent rule $Q_j^2 = Q_j$. This means that applying Q_j twice to any vector yields the same result as applying it once.

Using $Q_j = \eta_j \eta_j^T$, the solution can be re-expressed as:

$$S(t) = \sum_j e^{\gamma_j} Q_j S(0) \tag{36}$$

where the projection operator $Q_j = \eta_j \eta_j^T$ is incorporated into Sylvester's formula as follows:

$$Q_j = \prod_{j \neq k=1}^{N^2-1} \frac{G - \gamma_j I}{\gamma_k - \gamma_j} \tag{37}$$

Let $G(t_1)$ denote the integral $\int_0^t g(t_1')\, dt_1'$, where $g(t_1)$ is the matrix. Suppose γ_j are the eigenvalues of G. It can be shown that Sylvester's formula acts as a projection operator.

Proof: If G has n distinct eigenvalues, any vector $|\Psi\rangle$ in the n-dimensional space can be expanded in terms of the n eigenvectors

$$|\Psi\rangle = \sum_{k=1}^{n} \gamma_k |K\rangle \tag{38}$$

From the eigenvalue equation, we derive or obtain

$$G|j\rangle = \gamma_k |j\rangle \tag{39}$$

Therefore, to verify if Q_j, defined as $\prod_{j \neq k=1}^{N^2-1} \frac{G-\gamma_j I}{\gamma_k - \gamma_j}$, is a projection operator, we apply it to the state $|q\rangle$ as follows:

$$\begin{aligned} Q_j|q\rangle &= \prod_{j \neq k=1}^{N^2-1} \frac{G - \gamma_j I}{\gamma_k - \gamma_j} |q\rangle \\ Q_j|q\rangle &= \delta_{jq}|q\rangle \end{aligned} \tag{40}$$

where we used that $G|j\rangle = \gamma_k |j\rangle$

$$\begin{aligned} j \neq k, j = q, \quad & \left(G - \gamma_j I\right)|m\rangle = 0|q\rangle \\ j = k, \quad & \left(G - \gamma_j I\right)|m\rangle = 1|q\rangle \end{aligned} \tag{41}$$

Hence, an operator with eigenvalues 0 or 1 is a projection operator. It is worth noting here that $|m\rangle$ represents the eigensystem.

Let us revisit the example discussed in subsection (2.1) and apply Sylvester's formula to obtain the analytical solution of the coherence vector, or Bloch vector, for a two-level system. As before, we assume the system is initially prepared in the ground state. Initially, the equation of motion for the two-level system in terms of $SU(2)$ is:

$$\frac{d}{dt}\begin{pmatrix} S_1 \\ S_2 \\ S_3 \end{pmatrix} = \begin{pmatrix} 0 & \Delta & 0 \\ -\Delta & 0 & -\Omega \\ 0 & \Omega & 0 \end{pmatrix}\begin{pmatrix} S_1 \\ S_2 \\ S_3 \end{pmatrix} \tag{42}$$

its solution, assuming the matrix g is commutable with itself at different times, can be written as:

$$\vec{S}(t) = e^{G(t_1)}\vec{S}(0) \tag{43}$$

where $G(t_1) = \int_0^t g(t_1)dt_1$ and is expressible as:

$$\begin{aligned} G = \int_0^t g(t_1)dt_1 &= \int_0^t dt_1 \begin{pmatrix} 0 & \Delta & 0 \\ -\Delta & 0 & -\Omega \\ 0 & \Omega & 0 \end{pmatrix} \\ &= \begin{pmatrix} 0 & \int_0^t \Delta dt_1 & 0 \\ -\int_0^t \Delta dt_1 & 0 & -\int_0^t \Omega dt_1 \\ 0 & \int_0^t \Omega dt_1 & 0 \end{pmatrix} \\ &= \begin{pmatrix} 0 & \Delta' & 0 \\ -\Delta' & 0 & -\Omega' \\ 0 & \Omega' & 0 \end{pmatrix} \end{aligned} \tag{44}$$

where Δ' and Ω' represent the integrated value of Δ and Ω, respectively.

The next step is to obtain the eigensystems, that is, eigenvalues and eigenvectors, of g. Since our matrix is low dimensional we can find the eigenvalue by solving the characteristics equation $\det(g - \lambda I) = 0$ yielding eigenvalues of $\left\{0, -\sqrt{-\Delta'^2 - \Omega'^2}, \sqrt{-\Delta'^2 - \Omega'^2}\right\}$ and unnormalized eigenvectors

$$\eta = \begin{pmatrix} -\frac{\Omega'}{\Delta'} & \frac{\Delta'}{\Omega'} & \frac{\Delta'}{\Omega'} \\ 0 & -\frac{\sqrt{-\Delta'^2 - \Omega'^2}}{\Omega'} & \frac{\sqrt{-\Delta'^2 - \Omega'^2}}{\Omega'} \\ 1 & 1 & 1 \end{pmatrix} \tag{45}$$

Therefore, plugging the initial values of the coherence vector, $\vec{S}(0) = (0, 0, -1)^T$ which follows from definition of the vector $\vec{S} = (u_{01}, v_{01}, w_1)^T$, we find the following result for $\eta^{-1}S(0)$

$$\eta^{-1}S(0) = -\begin{pmatrix} \dfrac{\Delta'^2}{\Delta'^2+\Omega'^2} \\ \dfrac{\Omega'^2}{2\left(\Delta'^2+\Omega'^2\right)} \\ \dfrac{\Omega'^2}{2(\Delta'^2+\Omega'^2)} \end{pmatrix} \tag{46}$$

Therefore, our solution is now as follows:

$$\begin{aligned} S(t) &= \sum_j e^{\gamma_j}\eta_j\left(\eta_j^T S(0)\right) \\ S(t) &= e^{\gamma_1}\eta_1\left(\eta_1^T S(0)\right) + e^{\gamma_2}\eta_2\left(\eta_2^T S(0)\right) + e^{\gamma_3}\eta_3\left(\eta_3^T S(0)\right) \end{aligned} \tag{47}$$

with $\zeta = \sqrt{\Delta'^2 + \Omega'^2}$, noting that $\gamma_1 = 0$, $\gamma_2 = -i\zeta$ and $\gamma_3 = i\zeta$, we then obtain

$$S(t) = e^0\begin{pmatrix} -\dfrac{\Omega'}{\Delta'} \\ 0 \\ 1 \end{pmatrix}\left(\frac{\Delta'^2}{\zeta^2}\right) + e^{-i\zeta}\begin{pmatrix} -\dfrac{\Delta'}{\Omega'} \\ -\dfrac{-i\zeta}{\Omega'} \\ 1 \end{pmatrix}\left(\frac{\Omega'^2}{2\zeta^2}\right) + e^{i\zeta}\begin{pmatrix} \dfrac{\Delta'}{\Omega'} \\ \dfrac{-i\zeta}{\Omega'} \\ 1 \end{pmatrix}\left(-\frac{\Omega'^2}{2\zeta^2}\right) \tag{48}$$

which, after some algebraic manipulation, the solution becomes

$$\vec{S}(t) = \begin{pmatrix} \dfrac{\Delta'\Omega'}{\zeta^2}(1-\cos\zeta) \\ \dfrac{\Omega'}{\zeta}\sin\zeta \\ -\dfrac{\Delta'^2}{\zeta^2} - \dfrac{\Omega'^2}{\zeta^2}\cos\zeta \end{pmatrix} \tag{49}$$

Upon inspection, solution Eq. (49) reveals a form similar to that obtained with the adiabatic approximation, as shown in Eq. (14).

The plot in **Figure 2** concurs with the expectation mentioned earlier: the use of a laser pulse with an area of π completely transfers the population from the ground state to the excited state. Consequently, the state vector solution yields $S(t) = (0, 0, -1)^T$, with $w_1 = -1$ indicating complete population inversion.

2.2.2 Sylvester's formula for degenerate eigenvalues

So far, we have explored Sylvester's formula for systems with distinct eigenvalues. Now, let us consider systems with degenerate eigenvalues. In this context, a degenerate eigenvalue γ corresponds to two or more different linearly independent eigenvectors. Mathematically, this statement can be expressed as $GV_1 = \gamma V_1$ and $GV_2 = \gamma V_2$, where V_1 and V_2 are linearly independent eigenvectors.

The eigenvalues, in the context of this chapter, represent measurable values of physical observables, while the corresponding eigenstates represent the possible states in which the system may be found.

In such cases, the standard Sylvester's formula is not directly applicable to obtain the solution. This implies that when dealing with degenerate eigenvalues, modifications are necessary for Sylvester's formula to be used effectively to obtain the required solution. To address this, let m_j denote the algebraic multiplicity of the eigenvalue γ_j. The exponent in the solution $\vec{S}(t) = e^G \vec{S}(0)$ takes a modified form, as discussed in [12], which states:

$$e^G = \sum_j \left[\sum_{k=0}^{m_j - 1} b_k\left(\gamma_j\right)\left(G - \gamma_j I\right) \right] \prod_{j \neq i = 1} (G - \gamma_i I)^{m_j} \tag{50}$$

where the sum is performed over all distinct eigenvalue γ_j, and where the $b_k\left(\gamma_j\right)$ are the scalars

$$b_k\left(\gamma_j\right) = \frac{1}{k!} \frac{d^k}{d\gamma^k} \left[\frac{e^\gamma}{\prod_{j \neq i = 1}^{N^2 - 1} \left(\gamma - \gamma_j I\right)^{m_j}} \right]_{\gamma = \gamma_i} \tag{51}$$

Example - ii: Consider a three-level Λ system with states $\{|0\rangle, |1\rangle, |2\rangle\}$. In this system, there is no direct coupling between states $|0\rangle \leftrightarrow |2\rangle$, while levels are coupled by a coherent light with a half Rabi frequency $\alpha(t) = \frac{\hbar}{2}\Omega_p(t)$. Similarly, levels $|1\rangle \leftrightarrow |2\rangle$ are coupled by another coherent light with a half Rabi frequency $\beta(t) = \frac{\hbar}{2}\Omega_s(t)$, where $\alpha(t)$ and $\beta(t)$ are defined as the half Rabi frequencies of the pump and Stokes lights, respectively, see **Figure 3(a)**. The Hamiltonian of this system, which has been extensively studied [13–15], possesses three distinct eigenvalues. To seek an analytical solution using Sylvester's formula, we need to map the Hamiltonian into a larger space. One approach to achieve this is through the Lie algebraic method outlined in Ref. [16]. In this approach, physical observables are expressed in terms of generators of a Lie group, where the Hamiltonian and density matrix can be represented as linear combinations of these generators. It is shown that the Hamiltonian in this expanded space, denoted as $g_{\alpha\beta}$, takes the form of a matrix:

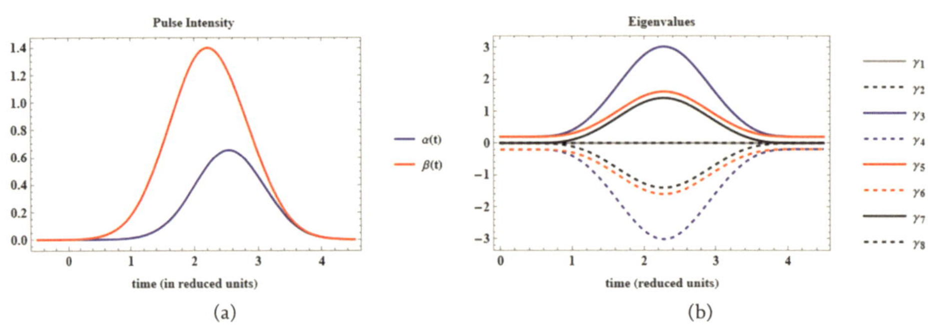

Figure 3.
(a) The pulse profile of the coupling lasers for the three-level system, showing the blue pump laser and the red Stokes laser. (b) The eight eigenvalues, with two of them being identical, specifically 0.

$$g = \begin{pmatrix} 0 & 0 & 0 & \Delta & 0 & \beta & 0 & 0 \\ 0 & 0 & 0 & 0 & -\Delta & -\alpha & 0 & 0 \\ 0 & 0 & 0 & \beta & -\alpha & 0 & 0 & 0 \\ -\Delta & 0 & -\beta & 0 & 0 & 0 & 2\alpha & 0 \\ 0 & \Delta & \alpha & 0 & 0 & 0 & -\beta & \sqrt{3}\beta \\ -\beta & \alpha & 0 & 0 & 0 & 0 & 0 & 0 \\ 0 & 0 & 0 & -2\alpha & \beta & 0 & 0 & 0 \\ 0 & 0 & 0 & 0 & -\sqrt{3}\beta & 0 & 0 & 0 \end{pmatrix} \tag{52}$$

The eigenvalues, γ_j, of this Hamiltonian are thus obtained to be as follows.

$$\begin{aligned} \gamma_{1,2} &= 0, \\ \gamma_{3,4} &= \pm\sqrt{-4\alpha^2 - 4\beta^2 - \Delta^2} \\ \gamma_{5,6,7,8} &= \pm\frac{\sqrt{-2\alpha^2 - 2\beta^2 - \Delta^2 \mp \Delta\sqrt{4\alpha^2 + 4\beta^2 - \Delta^2}}}{\sqrt{2}} \end{aligned} \tag{53}$$

The plots of these eigenvalues are shown in **Figure 3(b)**, and as we can see, two of the eigenvalues are zero, specifically $\gamma_1 = \gamma_2 = 0$, while the remaining six eigenvalues are purely imaginary. Therefore, due to the presence of degenerate eigenvalues, we need to use the second form of Sylvester's formula, denoted as Eq. (50). Here, G refers to the matrix obtained after integrating its elements, denoted as $G_2(t)$.

Assuming the system is initially prepared in the ground state with the initial condition of the vector $\vec{S}(0) = \left(0{,}0{,}0{,}0{,}0{,}0, -1, -\frac{1}{\sqrt{3}}\right)^T$, the evolution of the eight-dimensional coherence vector is then obtained using Sylvester's formula. The solution takes the form:

$$\begin{aligned} S(t) &= e^G S(0) \\ &= \left[(b_0(\gamma_1 = 0) + b_1(\gamma_2 = 0)(G - \gamma(=0)I_{8\times8})) \prod_{k\neq1,\,2k=3}^{8} (G - \gamma_k I_{8\times8}) + \sum_{j=3}^{8} e^{\gamma_j} \prod_{j\neq k=1}^{8} \frac{G - \gamma_k I_{8\times8}}{\gamma_j - \gamma_k} \right] \\ &\quad S(0) \end{aligned} \tag{54}$$

where $I_{8\times8}$ is an 8×8 identity matrix, and the scalar elements b_n (for $n = 0, 1$) corresponding to the two degenerate eigenvalues (with $\gamma_{1,2} = 0$) are expressed as

$$b_0(\gamma_i) = \frac{1}{0!}\frac{d^0}{d\gamma^0}\left[\frac{e^\gamma}{\prod_{k=3}^{8}(\gamma - \gamma_k I)}\right]_{\gamma=0} = \frac{1}{\prod_{k=3}^{8}(-\gamma_k)} \tag{55}$$

and

$$\begin{aligned} b_1(\gamma_i) &= \frac{1}{1!}\frac{d}{d\gamma}\left[\frac{e^\gamma}{\prod_{k=3}^{8}(\gamma - \gamma_k I)}\right]_{\gamma=0} \\ &= \frac{\prod_{k=3}^{8}(-\gamma_k I) - \prod_{k=3}^{8}(1 - \gamma_k I)}{\left[\prod_{k=3}^{8}(-\gamma_k I)\right]^2} \end{aligned} \tag{56}$$

3. Conclusions

In conclusion, we have explored analytical methods for solving the evolution of quantum systems with varying eigenvalue spectra. These analytical solutions are pivotal for gaining deep insights into system dynamics, facilitating precise predictions under different conditions, and validating theoretical models against experimental data. They also contribute to computational efficiency, offering rapid insights compared to numerical simulations, thereby advancing our understanding and utilization of quantum phenomena in diverse applications such as quantum information processing and quantum technologies. For the two-level case with distinct eigenvalues, the adiabatic approximation, Magnus solution, and Sylvester's formula provided a simplified yet effective method, yielding insightful predictions with clear physical interpretations. Sylvester's formula proved instrumental in analyzing the evolution of an eight-dimensional coherence vector in systems characterized by degenerate eigenvalues.

Author details

Dawit Hiluf Hailu
Department of Natural Sciences, Bowie State University, Bowie, USA

*Address all correspondence to: dhailu@bowiestate.edu

References

[1] Alhassid Y, Levine RD. Entropy and chemical change. III. The maximal entropy (subject to constraints) procedure as a dynamical theory. The Journal of Chemical Physics. 1977; **67**(10):4321-4339

[2] Hiluf D. Link between Alhassid-Levine and Hioe-Eberly Formalisms of $su(n)$ Equation of Motion. 2017. arXiv. Available from: https://arxiv.org/abs/1612.03276. Eprint: 1612.03276

[3] Hioe FT, Eberly JH. N-level coherence vector and higher conservation-laws in quantum optics and quantum-mechanics. Physical Review Letters. 1981;**47**(12):838-841

[4] Hailu D. Operation of CNOT Gate with Solutions Obtained Using 3rd Order Magnus Expansion. 2017. Available from: https://www.researchgate.net/publication/313820697_Operation_of_CNOT_gate_with_solutions_obtained_using_3_rd_order_Magnus_expansion

[5] Hailu DH. $su(2)$ Dynamics and Logic Machines – Part II. 2019. arXiv. Available from: https://arxiv.org/abs/1909.02094. Eprint: 1909.02094

[6] Magnus W. On the exponential solution of differential equations for a linear operator. Communications on Pure and Applied Mathematics. 1954; 7(4):649-673

[7] Masuo Suzuki. Generalized trotter's formula and systematic approximants of exponential operators and inner derivations with applications to many-body problems. Communications in Mathematical Physics. Jun 1976;51(2): 183-190

[8] Wei J, Norman E. Lie algebraic solution of linear differential equations. Journal of Mathematical Physics. 1963; **4**(4):575-581

[9] Wei J, Norman E. On global representations of the solutions of linear differential equations as a product of exponentials. Proceedings of the American Mathematical Society. 1964; **15**(2):327-334

[10] Wilcox RM. Exponential operators and parameter differentiation in quantum physics. Journal of Mathematical Physics. 1967;**8**(4):962-982

[11] Moler C, Van Loan C. Nineteen dubious ways to compute the exponential of a matrix, twenty-five years later. SIAM Review. 2003;**45**(1): 3-49

[12] Tarantola A. Elements for Physics: Quantities, Qualities, and Intrinsic Theories. 1st ed. Berlin, Heidelberg: Springer; 2006. pp. 266. DOI: 10.1007/978-3-540-31107-2

[13] Fewell MP, Shore BW, Bergmann K. Coherent population transfer among three states: Full algebraic solutions and the relevance of non adiabatic processes to transfer by delayed pulses. Australian Journal of Physics. 1997;**50**(2):281-308

[14] Shore BW. Coherent manipulations of atoms using laser light. Acta Physica Slovaca. 2008;**58**(3):243-486

[15] Vitanov NV, Stenholm S. Adiabatic population transfer via multiple intermediate states. Physical Review A. 1999;**60**(5):3820-3832

[16] Hiluf D. All optical programmable logic array (PLA). Journal of Physics Conference Series. 2018;**987**:012033

Chapter 4

Spectral Perturbation Theory of Hermitian Matrices

Marcus Carlsson

Abstract

While Hermitian Perturbation theory is significantly simpler in many aspects than the non-Hermitian one, explicit formulas for first- and second-order perturbations are often not present in standard reference works on the topic and hard to find elsewhere, especially in the presence of degenerate eigenvalues (i.e., of multiplicity higher than one). This chapter aims to fill in these gaps and also presents some new contributions, pertaining to degenerate eigenvalues. We focus on the local behavior of eigenvalues and eigenvectors as functions of *all* the matrix entries in the perturbation, giving concrete formulas for the gradient and Hessian of the corresponding multivariable Taylor series.

Keywords: Hermitian perturbation theory, Taylor series, Rayleigh-Schrödinger coefficients, Frechét derivatives of eigenvectors and values, perturbation of degenerate eigenvalues

1. Introduction

A matrix A is called self-adjoint, also known as Hermitian, if the matrix entries $A_{(k,l)}$ satisfy $A_{(k,l)} = \overline{A_{(l,k)}}$. Equivalently, an $n \times n$-matrix is self-adjoint if $\langle Ax, y\rangle = \langle x, Ay\rangle$ for every pair of vectors in $\mathbb{C}^n$, (using the Euclidean scalar product). Denote by $\mathcal{H}_n$ the set of such matrices and let $A \in \mathcal{H}_n$ be fixed. The dependence of the eigenvalues of A, given a "small" perturbation $E \in \mathcal{H}_n$, is a classical subject with original contributions dating back to Lord Rayleigh in the nineteenth century [1]. The literature on this topic is immense and can roughly be divided into two groups. One group "freezes" the matrix E and considers $A + tE$ as a function of the complex variable t, giving rise to a beautiful and rich connection with algebra and analytic function theory [2, 3]. However, it lacks a global perspective, in the sense that E is fixed and not a free variable. The second group of results do not freeze E, with more general but less exact results as a consequence, such as the estimates by Geršgorin, Weyl, Stewart, and Bauer-Fike to name a few. In particular, if we denote the eigenvalues of A by $\alpha \in \mathbb{R}^n$ and those of $A + E$ by $\xi \in \mathbb{R}^n$ (both ordered non-increasingly), then the estimates of H. Weyl imply that

$$|\xi_j - \alpha_j| \leq \|E\|, \tag{1}$$

where the norm refers to the standard operator norm, that is,

$$\|A\| = \sup\{\|Ax\|/\|x\| : x \neq 0\} = \max\left\{|\alpha_j|\right\}_{j=1}^{n}.$$

While this result is the best possible from a global perspective, it offers little guidance on how ξ depends on E when $\|E\|$ is small, and it is the key objective of this chapter to provide concrete and constructive formulas filling this gap.

To begin, let us first consider perturbations of the form $A + tE$ for a fixed $E \in \mathcal{H}_n$. A key result in this field is the fact that the eigenvalues $\xi(t)$ of $A + tE$ are real analytic functions of $t \in \mathbb{R}$ (given a suitable ordering). This result is due to F. Rellich in a sequence of articles from the 30's [4], and a simple proof in the finite dimensional setting is found in his monograph [5] (or consult Theorem 6.1 Ch. II in [3]). Even before that, the coefficients of the corresponding series expansion were computed by Lord Rayleigh and later Schrödinger, although they lacked a general proof that the corresponding series converged. These coefficients are typically found in the literature on mathematical physics and quantum physics, rather than books on pure mathematics such as Kato's seminal work [3] or Rellich's own monograph [5], for that matter. For example, they are computed in Reed-Simon's book [6], Ch. XII.1, using complex analytic tools, while assuming that the eigenvalues of A are simple. Even without this assumption, Courant and Hilbert [7] compute them by making a simple "ansatz" and backing out their values from a set of equation systems. While these coefficients have very complicated expressions, the first- and second-order terms are manageable; if we suppose for simplicity that a basis has been chosen such that $A = \Lambda_\alpha$, where Λ_α denotes the corresponding diagonal matrix

$$\Lambda_\alpha = \begin{pmatrix} \alpha_1 & 0 & \dots & 0 \\ 0 & \alpha_2 & \ddots & \vdots \\ \vdots & \ddots & \ddots & \\ 0 & \dots & & \alpha_n \end{pmatrix},$$

and moreover such that $E(i,j) = 0$ whenever $\alpha_i = \alpha_j$ and $i \neq j$, then

$$\xi_k(t) = \alpha_k + tE(k,k) + t^2 \sum_{l:\alpha_l \neq \alpha_k} \frac{|E(k,l)|^2}{\alpha_k - \alpha_l} + O(t^3). \tag{2}$$

The formula is quite accurate in practice, see **Figure 1** which displays a histogram of the errors $\xi_k(t) - \left(\alpha_k + tE(k,k) + t^2 \sum_{l:\alpha_l \neq \alpha_k} \frac{|E(k,l)|^2}{\alpha_k - \alpha_l}\right), k = 1 \dots n$, for five randomly generated instances with $n = 30$. For this realization, we set $t = 10^{-3}$, and it is noteworthy that all errors are well below 10^{-6}. The outliers of magnitude 10^{-7} are caused by eigenvalues of the unperturbed matrix that are very close, in which case the formula (2) becomes less precise, highlighting the fact that degenerate eigenvalues are a major obstacle, in both theory and practice.

Despite the beauty of formula (2) (which is proved in Section 5), it lacks a global perspective. In his 1953 monograph, Rellich himself points out that even introducing two unknown parameters in the perturbation leads to lack of analyticity and unpredictable behavior. In this chapter, our main focus is to prove concrete results of the above type with E as a free variable. We shall also show that, in the case when A

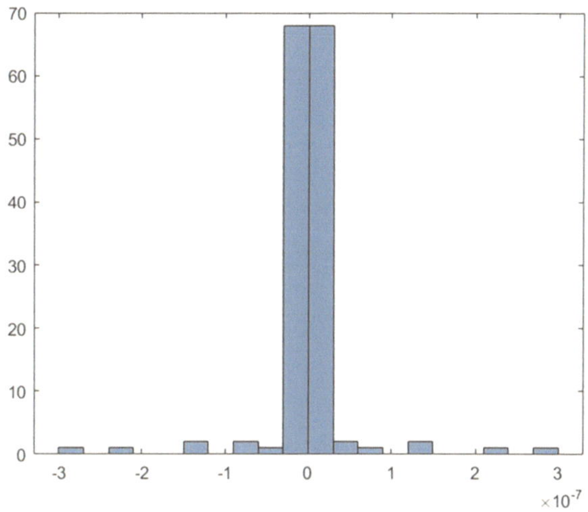

Figure 1.
Histogram of the error in the approximation (2) for five randomly generated examples with $n = 30$.

has distinct eigenvalues, the perturbed eigenvalues and eigenvectors depend analytically on E and provide formulas for their respective Frechét derivatives as well as corresponding second-order terms (i.e., the Hessian as a bilinear vector-valued function). We also consider the more difficult case when A has degenerate eigenvalues.

In order to say anything concrete about the perturbed eigenvalues and eigenvectors, it is of course necessary to know the eigenvectors and eigenvalues of A to begin with. It is furthermore no restriction to assume that the eigenvectors are orthogonal and normalized, that is, we suppose that a unitary matrix U_A is known such that $A = U_A \Lambda_\alpha U_A^*$. Given any perturbation E, the operation $E \mapsto \hat{E} := U_A^* E U_A$ is then a linear isometry and $A + E = U_A(\Lambda_\alpha + \hat{E})U_A^*$. Hence, any perturbation theory result about $\Lambda_\alpha + \hat{E}$, whether it deals with the eigenvectors or the eigenvalues, can easily be lifted to the more general situation considering $A + E$. Therefore, in order to keep notation simple, we shall always assume from the outset that $A = \Lambda_\alpha$ and that $\hat{E} = E$.

2. The case of distinct eigenvalues: Real analyticity and Frechét-differentiability

From an applied point of view, the by far most common situation is that the eigenvalues α are distinct, and in this case, the theory is much simpler than in the general case. We therefore consider this case first and postpone a study of the more general situation to Section 4. Thus, we shall in this section and the next assume that $\alpha_i \neq \alpha_j$ whenever $i \neq j$ and, for concreteness, we always assume that both α and ξ are ordered decreasingly, where ξ denotes the eigenvalues of $\Lambda_\alpha + E$ for any perturbation $E \in \mathcal{H}_n$.

Note that $\mathcal{H}_n$ is a linear space of real dimension n^2 which can be parameterized, for example, by the isometric isomorphism

$$\mathbb{R}^{n^2} \ni (a_1, \dots, a_n, b_1, \dots, b_{(n^2-n)/2}, c_1, \dots, c_{(n^2-n)/2}) \mapsto$$

$$\iota(a,b,c) := \begin{pmatrix} a_1 & \frac{b_1 + ic_1}{\sqrt{2}} & \frac{b_2 + ic_2}{\sqrt{2}} & \cdots & \frac{b_{n-1} + ic_{n-1}}{\sqrt{2}} \\ \frac{b_1 - ic_1}{\sqrt{2}} & a_2 & \frac{b_n + ic_n}{\sqrt{2}} & \cdots & \frac{b_{2n-3} + ic_{2n-3}}{\sqrt{2}} \\ \frac{b_2 - ic_2}{\sqrt{2}} & \frac{b_n - ic_n}{\sqrt{2}} & a_3 & \cdots & \frac{b_{3n-6} + ic_{3n-6}}{\sqrt{2}} \\ \vdots & \vdots & \vdots & \ddots & \vdots \\ \frac{b_{n-1} - ic_{n-1}}{\sqrt{2}} & \frac{b_{2n-3} - ic_{2n-3}}{\sqrt{2}} & \frac{b_{3n-6} - ic_{3n-6}}{\sqrt{2}} & \cdots & a_n \end{pmatrix}.$$

We can therefore consider the perturbed eigenvalues $\xi(E)$ as multivariable functions on $\mathbb{R}^{n^2}$ as well as functions defined on the linear space $\mathcal{H}_n$. Both viewpoints turn out to be instructive. We shall prove that, from the multivariable perspective the eigenvalues are real analytic functions and that, from the linear space perspective, these are Frechét differentiable and provide a concrete formula without involving any partial derivatives. To be precise, we shall say that "ξ is real analytic in the matrix coefficients of E" whenever the function $\xi \circ \iota$ is real analytic on $\mathbb{R}^{n^2}$ in a neighborhood of 0 (where $\xi \circ \iota$ denotes the composition $(a,b,c) \mapsto \xi(\iota(a,b,c))$).

Before presenting these results, let us also discuss perturbation of the eigenvectors. Unfortunately, the eigenvectors are not unique and therefore, it is not immediately clear how to define $U(E)$ in a way such that $U(E)$ contain a collection of perturbed eigenvectors. The most natural choice for $U(0)$, that is, the eigenvectors of Λ_α, is clearly the identity matrix I, so therefore one natural rule for uniquely determining the eigenvectors (in a neighborhood of 0) is to demand that $\mathtt{diag}(U(E)) = (1, \dots, 1)$, where $\mathtt{diag} : \mathbb{C}^{n \times n} \to \mathbb{C}^n$ is the operator that extracts the diagonal values of any matrix. This rule is computationally convenient, but unfortunately does not entail that $U(E)$ becomes unitary. However, this is easily remedied by post-normalizing the columns. We therefore let $V(E)$ denote the matrix containing the eigenvectors to $\Lambda_\alpha + E$ that satisfy $\mathtt{diag}(V(E)) = (1, \dots, 1)$, and we define the unitary eigenvector-matrix $U(E)$ by setting $u_j = v_j / \|v_j\|$, where u_j and v_j denote the jth column of $U(E)$ and $V(E)$ respectively. Note that $U(E)$ indeed becomes a unitary matrix, since orthogonality of the columns is automatic by the spectral theorem. We summarize the above in the below definition.

Definition 2.1. *Given a fixed matrix Λ_α and a "small perturbation" E, we denote by $\xi(E)$ the (decreasingly ordered) eigenvalues of $\Lambda_\alpha + E$, and we denote by $U(E)$ the unitary eigenvector-matrix as defined above.*

By "small perturbation" we mean that E is confined to a ball around 0 in $\mathcal{H}_n$ such that there are no occurrences of degenerate eigenvalues, which is possible since we have assumed that Λ_α has distinct eigenvalues. Our first result can be seen as a multivariate extension of Rellich's theorem.

Theorem 2.2. *The functions ξ and U are real analytic in the matrix coefficients of E.*

Proof. The eigenvalues $\xi(\iota(a,b,c))$ are the λ-roots of the polynomial $p : \mathbb{C} \times \mathbb{R}^{n^2} \to \mathbb{R}$ defined by $p(\lambda, a, b, c) = \det(\Lambda_\alpha + \iota(a,b,c) - \lambda I)$. Standard residue calculus thus gives that

$$\xi_j(E) = \int_{\Gamma_j} \frac{\zeta}{p(\zeta, a, b, c)} \frac{d\zeta}{2\pi i}$$

for small enough coefficients (a, b, c), where Γ_j is a small circle centered around α_j. If we allow the coefficients to take complex values, then the integrand $\frac{\zeta}{p(\zeta, a, b, c)}$, for fixed $\zeta \in \Gamma_j$, becomes holomorphic in each of its arguments a_j, b_k, and c_l, respectively, where $1 \le j \le n$ and $1 \le k, l \le (n^2 - n)/2$. In other words we have that

$$\bar{\partial}_x \frac{\zeta}{p(\zeta, a, b, c)} = 0$$

holds for each x representing either of the variables a_j, b_k, or c_l, where $\bar{\partial}_x$ represents the famous "d-bar" differential operator. Standard facts from real analysis then show that $\bar{\partial}_x$ and $\int_{\Gamma_j}$ commute, which implies that $\bar{\partial}_x \xi_j = 0$ for each x (representing a_j, b_k, or c_l). It now follows from Harthog's theorem that ξ_j is analytic as a function of several variables, in a ball around 0 in $\mathbb{C}^{n^2}$. The restriction of this function to $\mathbb{R}^{n^2}$ is then by definition real analytic, as was to be shown.

We now tackle the corresponding result for the eigenvectors, and begin by proving analyticity for the eigenvector matrix V discussed before Definition 2.1. By the above we have that the column-vector $v_j(a, b, c)$ of $V(\imath(a, b, c))$ solves the equation

$$\left(\Lambda_\alpha + E - \xi_j I\right) v_j = 0$$

where for ease of reading we let the dependence on (a, b, c) be implicit, that is, we abbreviated $\imath(a, b, c)$ and $\xi(\imath(a, b, c))$ by E and ξ, respectively. The matrix $\Lambda_\alpha + E - \xi_j I$ is, by the first part of the proof, real analytic in (a, b, c). Let $X_j = X_j(a, b, c)$ denote the same matrix but with the j'th row replaced by the canonical basis vector e_j (i.e., the vector which is 0 on all positions except the j'th, where it equals 1). For $(a, b, c) = (0,0,0)$ it is clearly invertible, and hence, it is also invertible in a neighborhood of 0, by continuity. Since we have assumed the normalization $v_j(j) = 1$, it follows that v_j solves the equation $X_j v_j = e_j^T$, where T denotes the transpose. This equation can be solved explicitly for v_j using Cramer's rule, which only involves basic operations that preserve real analyticity, and hence, the real analyticity of X_j implies the real analyticity of v_j, which was to be shown.

Finally, since real analyticity of a complex-valued function implies that the real and imaginary part are real analytic by themselves, and since we can write

$$u_j = v_j / \sqrt{\sum_{k=1}^{n} (\mathrm{Re}\, v_j(k))^2 + (\mathrm{Im}\, v_j(k))^2},$$

it follows that the columns of $U(E)$ are real analytic as well. □

As a consequence of the above result we see that $\xi \circ \imath$ and $U \circ \imath$ are C^∞-functions, so Taylor's formula implies that

$$\xi(\imath(a, b, c)) = \xi(0) + \langle \nabla \xi \circ \imath|_0, (a, b, c) \rangle + O\left(\|(a, b, c)\|^2\right), \tag{3}$$

and similarly for higher-order expansions. Translated back to $\mathcal{H}_n$ via ι^{-1} (which is isometric so preserves norms and scalar products by the polarization identity), (3) means that $E \mapsto \xi(E)$ is Frechét differentiable, that is, that there exists some linear operator $L_\xi : \mathcal{H}_n \to \mathbb{R}^n$ (the Frechét-derivative of ξ) such that

$$\xi(E) = \alpha + L_\xi(E) + O\left(\|E\|^2\right), \tag{4}$$

where we used that $\xi(0) = \alpha$. To be more precise, we have that $L_\xi(E) = \langle \iota^{-1}(\nabla \xi \circ \iota|_0), E \rangle$. However, the value of Theorem 2.2 is not so much the real analyticity in itself, at least from an applied perspective, but rather the fact that there exists a linear operator L_ξ such that (4) holds. Indeed, finding $L_\xi(E)$ based on the formula $\iota^{-1}(\nabla \xi \circ \iota|_0)$ and computing partial derivatives of $\xi \circ \iota$ would be a nightmare. Instead, just knowing that L_ξ exists, we can now find a formula for $L_\xi(E)$ which involves only basic (and efficiently computable) matrix operations on E directly.

To do so, first note that by applying the same argument as above to each of the coordinate functions in $U(E)$, we also have

$$U(E) = I + L_U(E) + O\left(\|E\|^2\right) \tag{5}$$

where L_U is a linear operator from $\mathcal{H}_n$ to $\mathbb{C}^{n \times n}$ (and we used that $U(0) = I$). By plugging these expressions into the defining equation for ξ and U, that is,

$$U(E)\Lambda_{\xi(E)} = (\Lambda_\alpha + E)U(E), \tag{6}$$

we can easily find the sought formulas for both L_ξ and L_U. To this end, we introduce the matrix $M \in \mathbb{R}^{n \times n}$ be defined by

$$M(j,k) = \left(\alpha_k - \alpha_j\right)^{-1} \tag{7}$$

for $j \neq k$ and 0 on the diagonal. We then have:

Theorem 2.3. *Both ξ and U are Frechét differentiable at 0, and $L_\xi(E) = \mathtt{diag}(E)$ whereas $L_U(E) = M \diamond E$ where $\diamond$ represents Hadamard matrix-multiplication.*

Proof. The Frechét differentiability has already been established above, so let us focus on obtaining the acclaimed formulas for L_ξ and L_U. It will be somewhat easier to first derive formulas for ξ and the matrix V, used in the construction of U, and then use these to derive the sought formulas for U itself. Upon inserting (4) and (5) (or rather the counterpart for V) in (6), we obtain the equation

$$(I + L_V(E))\left(\Lambda_\alpha + \Lambda_{L_\xi(E)}\right) = (\Lambda_\alpha + E)(I + L_V(E)) + O\left(\|E\|^2\right).$$

Upon ignoring all but the terms that are linear in E, this gives

$$[L_V(E), \Lambda_\alpha] + \Lambda_{L_\xi(E)} = E, \tag{8}$$

where $[L_V(E), \Lambda_\alpha]$ denotes the commutator $L_V(E)\Lambda_\alpha - \Lambda_\alpha L_V(E)$. Now, irrespective of what $L_V(E)$ equals, the commutator is by construction 0 on the diagonal. Hence, solving for L_ξ we obtain $L_\xi(E) = (E(1,1), \ldots, E(n,n)) = \mathtt{diag}(E)$, as desired. Looking at the off-diagonal elements we similarly get the equation

$(L_V(E)(j,k))(\alpha_k - \alpha_j) = E(j,k)$,[1] implying $L_V(E)(j,k) = E(j,k)/(\alpha_k - \alpha_j)$ whenever $j \neq k$. Moreover, by the normalization used for the matrix V we have $V(E)(j,j) \equiv 1$ which yields that $L_V(E)$ must be 0 on the diagonal. Summing up, we have established that

$$L_V(E) = M \diamond E. \tag{9}$$

We now derive the corresponding formula for L_U. If, as before, we let v_j denote the columns of V, then $U = V\Lambda_{\left(\|v_j\|^{-1}\right)_{j=1}^{n}}$. Since $v_j(l) = V(l,j)$ we derive that

$$\|v_j\|^{-1} = \left(\sqrt{1 + \sum_{l:l\neq j} |V(l,j)|^2}\right)^{-1} = \left(\sqrt{1 + \sum_{l:l\neq j} \left|\frac{E(l,j)}{\alpha_j - \alpha_l}\right|^2 + O(\|E\|^3)}\right)^{-1}$$

where the last line follows from the already established fact that $V = I + M \diamond E + O(\|E\|^2)$. Since $\left(\sqrt{1+x}\right)^{-1} = 1 - \frac{1}{2}x + O(x^2)$, it follows that

$$\|v_j\|^{-1} = 1 - \frac{1}{2}\sum_{l:l\neq j} \left|\frac{E(l,j)}{\alpha_j - \alpha_l}\right|^2 + O(\|E\|^3). \tag{10}$$

In summary, we have

$$U = V\Lambda_{\left(\|v_j\|^{-1}\right)_{j=1}^{n}} = \left(I + L_V(E) + O(\|E\|^2)\right)\left(I + O(\|E\|^2)\right) = I + M \diamond E + O(\|E\|^2), \tag{11}$$

which proves that $L_U(E) = M \diamond E$, as desired. □

Theorem 2.3 is illustrated in **Figure 2**. It is noteworthy that the estimate (4) for ξ consistently outperforms the estimate (5) for U, and by a large margin, 10^{-9} versus 10^{-6}. To understand why this is so, note by looking at the red plot of subplot 1 that α

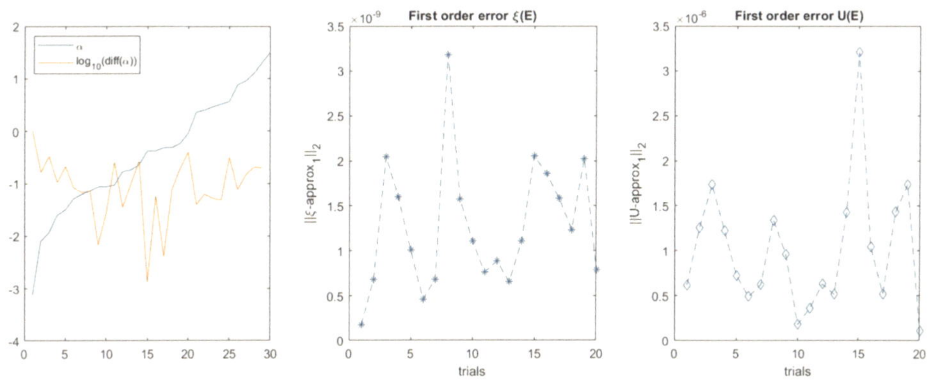

Figure 2.
Illustration of Theorem 2.3. Left figure displays the randomly generated eigenvalues of Λ_α, $n = 30$, along with a $\log_{10}$-plot of the difference of adjacent eigenvalues. The middle graph shows the error of formula (4) (ignoring the ordo-term) in ℓ^2-norm (i.e., $\|\xi(E) - \alpha - L_\xi(E)\|_2$), for 20 realizations of perturbations E with $\|E\| = 10^{-5}$. The right plot shows the corresponding graph for the corresponding error in (5), measured using the Frobenius norm.

[1] Lacking a better notation, we use $L_V(E)(j,k)$ to denote the (j,k)'th element of the matrix $L_V(E)$.

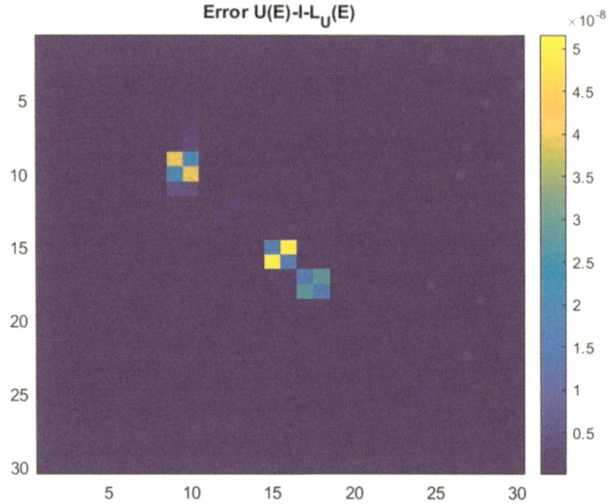

Figure 3.
*Modulus of the error $U(E) - I - L_U(E)$ in (5) for a given random E and same A as in **Figure 2**.*

has eigenvalues that are close to each other at indices $(9, 10)$, at indices $(15, 16)$ as well as at $(17, 18)$. **Figure 3** displays the modulus of the elementwise error in (5), using the same A as in **Figure 2** and one random realization of E as in **Figure 2** (i.e., normalized so that $\|E\| = 10^{-5}$). As is plain to see, most entries are of the order 10^{-9} as well, except for a few which corresponds precisely to blocks of indices j, k for which $\alpha_j \approx \alpha_k$ (i.e., $(9, 10)$, $(15, 16)$, and $(17, 18)$). This illustrates that the behavior of the eigenvectors is particularly unstable near semi-degenerate eigenvalues, but that this instability is confined to the above mentioned blocks. Sections 4 and 5 have more on this. Before that, lets improve the accuracy by finding the second-order terms in the Taylor expansions of ξ and U.

3. The case of distinct eigenvalues: Higher-order formulas

The above argument can now be bootstrapped to yield Taylor terms of arbitrarily high order. To generalize the notion of Hessian, we say that a function H, depending on two variables $(E, F) \in \mathcal{H}_n^2$, is real bilinear if it is linear in both arguments separately when we view $\mathcal{H}_n$ as a linear space over $\mathbb{R}$. In other words, we demand that $H(aA + bB, C) = aH(A, C) + bH(B, C)$ for all $a, b \in \mathbb{R}$ and $A, B \in \mathcal{H}_n$, and analogously for the second variable. Here, the target space will be either $\mathbb{C}^n$ of $\mathbb{C}^{n\times n}$. Also, H is said to be symmetric if $H(A, B) = H(B, A)$.

Theorem 3.1. *There exists a unique symmetric real bilinear function $H_\xi : \mathcal{H}_n^2 \to \mathbb{C}^n$ such that*

$$\xi(E) = \alpha + L_\xi(E) + \frac{1}{2}\mathcal{H}_\xi(E, E) + O(\|E\|^3), \tag{12}$$

which is given by the formula

$$H_\xi(E, F) = \left(\sum_{l:l\neq j} \frac{E(j,l)\overline{F(j,l)} + F(j,l)\overline{E(j,l)}}{\alpha_j - \alpha_l} \right)_{j=1}^{n}. \tag{13}$$

Remarks:

1. In particular, we have that

$$H_\xi(E,E) = 2\left(\sum_{l:l\neq j} \frac{|E(j,l)|^2}{\alpha_j - \alpha_l}\right)_{j=1}^{n}. \tag{14}$$

2. Formula (13) clearly boils down to (2) if we insert tE in place of E and consider E to be fixed and $t \in \mathbb{R}$ to be variable. While it is technically possible to prove (13) based on (2), we prefer the below self-contained proof, since it follows the lines Hilbert and Courant use to establish (2) and thus, it makes little sense to first prove (2) and then "lift" this result to (13) via an additional argument.

Proof. Adding the second-order term in the Taylor expansion (3) we see that

$$\xi(\iota(a,b,c)) = \xi(0) + \langle \nabla\xi\circ\iota|_0, (a,b,c)\rangle + \frac{1}{2}\langle \nabla^2(\xi\circ\iota)|_0 (a,b,c)^T (a,b,c)^T\rangle + O(\|(a,b,c)\|^3), \tag{15}$$

where $\nabla^2(\xi\circ\iota)\big|_0$ denotes the Hessian matrix at 0 for $\xi\circ\iota$. This proves that (12) holds with the real bilinear symmetric function H_ξ given by

$$H_\xi(E,F) = \langle \nabla^2(\xi\circ\iota)\big|_0 \left(\iota^{-1}(E)\right)^T, \left(\iota^{-1}(F)\right)^T\rangle,$$

but, in order to avoid a computational mayhem, we will not use this formula for the derivation of the explicit expression (13). Instead, we note that a similar argument applied to $V\circ\iota$ gives[2] that there exists a real bilinear symmetric function $H_V : \mathcal{H}_n^2 \to \mathbb{C}^{n\times n}$ such that

$$V(E) = I + L_V(E) + \frac{1}{2}H_V(E,E) + O(\|E\|^3). \tag{16}$$

Armed with this, we now play the same game as in the previous section. Inserting (12) and (16) in (6) we get the equation

$$\left(I + L_V + \frac{1}{2}H_V\right)\left(\Lambda_\alpha + \Lambda_{L_\xi} + \frac{1}{2}\Lambda_{H_\xi}\right) = (\Lambda_\alpha + E)\left(I + L_V + \frac{1}{2}H_V\right) + O(\|E\|^3),$$

where we suppressed the dependence on E for readability. By the results in the previous section the constant and linear terms cancel out. Also, the uniqueness of Taylor expansions imply that also the second-order terms (in E) must be identical, which gives the equation

$$\frac{1}{2}H_V\Lambda_\alpha + L_V\Lambda_{L_\xi} + \frac{1}{2}\Lambda_{H_\xi} = EL_V + \frac{1}{2}\Lambda_\alpha H_V \tag{17}$$

[2] As in the proof of Theorem 2.3 it is more convenient to work with V, introduced before Definition 2.1, than with U directly.

which, upon reshuffling and recalling (9), yields

$$[H_V, \Lambda_\alpha] + \Lambda_{H_\xi} = 2(EL_V - L_V\Lambda_{L_\xi}) = 2\Big(E(M\diamond E) - (M\diamond E)\Lambda_{\mathtt{diag}(E)}\Big). \tag{18}$$

Since both the commutator and the matrix $(M\diamond E)\Lambda_{\mathtt{diag}(E)}$ are zero on their diagonals, this gives the equation $H_\xi = 2\mathtt{diag}(E(M\diamond E))$, which written out reads

$$(E(M\diamond E))(j,j) = \sum_{l:l\neq j} E(j,l)\frac{E(l,j)}{\alpha_j - \alpha_l} = \sum_{l:l\neq j}\frac{|E(j,l)|^2}{\alpha_j - \alpha_l} \tag{19}$$

where we used that $E(l,j) = \overline{E(j,l)}$. This proves (14). Since the function (13), for each fixed l, is a symmetric real bilinear function which coincides with (14) when restricted to the diagonal (E,E), we are done. □

With a bit of more work we can also solve for H_V and eventually also for H_U.

Theorem 3.2. *There exists a unique symmetric real bilinear function $H_U : \mathcal{H}_n^2 \to \mathbb{C}^{n\times n}$ such that*

$$U(E) = I + L_U(E) + \frac{1}{2}H_U(E,E) + O(\|E\|^3), \tag{20}$$

which is given by the formula

$$H_U(E,F)(j,k) = \begin{cases} -\displaystyle\sum_{l\neq j}\frac{E(j,l)\overline{F(j,l)}}{(\alpha_l - \alpha_j)^2}, & j = k \\ \left(\displaystyle\sum_{l:l\neq k}\frac{E(j,l)\overline{F(k,l)} + F(j,l)\overline{E(k,l)}}{(\alpha_k - \alpha_l)(\alpha_k - \alpha_j)}\right) - \dfrac{E(j,k)F(k,k) + F(j,k)E(k,k)}{(\alpha_k - \alpha_j)^2}, & j\neq k \end{cases} \tag{21}$$

Proof. We now consider the off-diagonal terms in (18). As in (19) we get

$$2(E(M\diamond E))(j,k) = \sum_{l:l\neq k} E(j,l)\frac{E(l,k)}{\alpha_k - \alpha_l} = 2\sum_{l:l\neq k}\frac{E(j,l)\overline{E(k,l)}}{\alpha_k - \alpha_l} \tag{22}$$

but now we also have to take the term

$$\Big(-2(M\diamond E)\Lambda_{\mathtt{diag}(E)}\Big)(j,k) = -2\frac{E(j,k)}{\alpha_k - \alpha_j}E(k,k)$$

into account. Since this equals $([H_V, \Lambda_\alpha])(j,k) = (\alpha_k - \alpha_j)(H_V)(j,k)$ (where H_V is short for $H_V(E,E)$) we obtain the expression

$$H_V(j,k) = 2\left(\sum_{l:l\neq k}\frac{E(j,l)\overline{E(k,l)}}{(\alpha_k - \alpha_l)(\alpha_k - \alpha_j)}\right) - 2\frac{E(j,k)E(k,k)}{(\alpha_k - \alpha_j)^2}. \tag{23}$$

Since $V(j,j)$ is constant, we obtain an expression for $H_V(E,E)$ by defining it to be 0 on the diagonal and by the above equation for off-diagonal entries.

Now, to find the corresponding formula for $H_U(E,E)$ we need to normalize the columns v_j of V by $\|v_j\|$, as we did in the proof of Theorem 2.3. If we let η denote the vector that contain the second-order contributions from $\|v_j\|^{-1}$, that is,

$$\eta_j(E) = -\sum_{l:l\neq j}\left|\frac{E(l,j)}{\alpha_j-\alpha_l}\right|^2 = -\sum_{l:l\neq j}\frac{|E(j,l)|^2}{(\alpha_l-\alpha_j)^2} \tag{24}$$

where we used (10) and the fact that $E^* = E$, we see as in (11) that

$$U(E) = V(E)\Lambda_{\left(\|v_j(E)\|^{-1}\right)_{j=1}^{n}} =$$
$$\left(I + L_V(E) + \frac{1}{2}H_V(E) + O(\|E\|^3)\right)\left(I + \frac{1}{2}\Lambda_\eta + O(\|E\|^3)\right) =$$
$$I + M\diamond E + \frac{1}{2}\left(H_V(E,E) + \Lambda_{\eta(E)}\right) + O(\|E\|^3),$$

It follows that $H_U(E,E) = H_V(E,E) + \Lambda_{\eta(E)}$, and it is easy to see from (23) and (24) that the expression $H_U(E,F)$ given in the statement of the theorem is a symmetric real bilinear function which coincides with this on the "diagonal" $H_U(E,E)$. By uniqueness of such functions, the proof is complete.

□

Continuing in this manner, one can inductively retrieve formulas for the higher-order terms in the Taylor expansion. However, since these are rarely used in practice, and since the author has failed to find any simple structure for the general case, we stop here.

The efficacy of Theorems 3.1 and 3.2 is tested in **Figure 4**, where we run the same test as in **Figure 2** but including H_ξ and H_U, respectively, in the approximate formulas. We see that both the error for ξ and U drops by a factor of 10^{-3}.

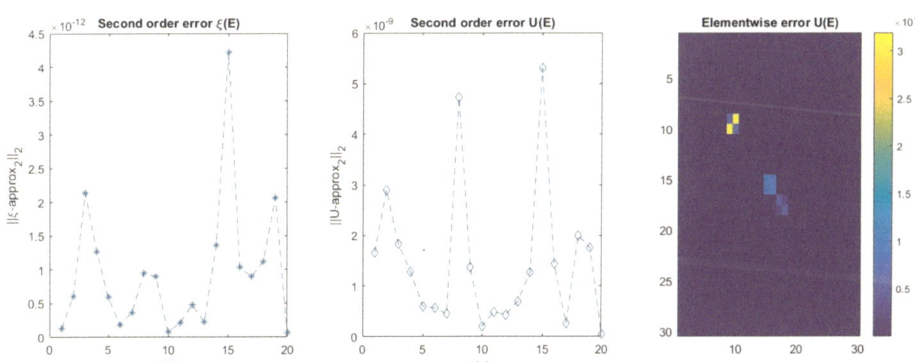

Figure 4.
*Illustration of Theorems 3.1 and 3.2, using the same example as in **Figure 2** (with the same random matrices). The left and middle plot now shows the error of formulas (12) and (20) when the ordo term is ignored (in ℓ^2-norm and Frobenius norm, respectively) for the same randomly generated E's as in **Figure 2**, where the corresponding first-order formulas (4) and (5) were used in the approximations. The right plot is the same as in **Figure 3** (with same E), where the only difference is that we now approximate using (20) as opposed to (5).*

4. Perturbation of degenerate eigenvalues

Following Reed and Simon [6], we shall say that an eigenvalue is degenerate if it has multiplicity $m > 1$. First of all, in this situation the perturbed eigenvalues are *not* real analytic in the matrix coefficients, in stark contrast to Rellich's theorem which states that the eigenvalues (and vectors) of a "line-perturbation" $A + tE$ always are real analytic in t, independent of whether A has distinct eigenvalues or not. To see this it suffices to consider the 2x2-case and pick A to be the zero matrix. With

$$E = \iota(a_1, a_2, b, c) = \begin{pmatrix} a_1 & (b+ic)/\sqrt{2} \\ (b-ic)/\sqrt{2} & a_2 \end{pmatrix}$$

we then have that the eigenvalues equal $\frac{a_1+a_2}{2} \pm \sqrt{\frac{(a_1+a_2)^2}{4} + \frac{b^2+c^2}{2}}$. Upon computing the gradient of this expression we see that the eigenvalues are not even C^1. They are, however, Lipschitz continuous, as follows from Weyl's inequality (1). To make this chapter self-contained, we provide a basic proof of this fact.

Theorem 4.1. *Let $A \in \mathcal{H}_n$ be arbitrary and denote by ξ the eigenvalues of $A + E$, ordered non-increasingly, where $E \in \mathcal{H}_n$ is a variable. Then each ξ_j, $1 \le j \le n$, is Lipshitz continuous with constant 1 (with respect to the operator norm).*

We remark that, since the eigenvectors of $A + E$ in the above 2x2 example coincide with those of E, which can be any orthonormal pair, we see that the eigenvectors are not even continuous, so there is little hope of saying anything similar about the eigenvectors.

Proof. The eigenvalues are the roots of the equation $\det(\Lambda_\alpha + \iota(a,b,c) - \lambda I) = 0$, and it is a standard fact from basic algebra that the roots of a polynomial equation depend continuously on the polynomial coefficients, which in turn depend continuously on the matrix entries (a,b,c). To see that they are also Lipschitz with constant 1, fix j, and fix some $\epsilon > 0$. We may assume that ε is small enough that it is less than half of the *isolation distance* of α_j; that is, $\min_{\alpha_l \ne \alpha_j}\{|\alpha_l - \alpha_j|\}$. Let Γ_j be the circle with center α_j and radius ϵ. In order for one of the eigenvalues that are inside Γ_j (when $E = 0$) to escape to the outside when E varies, we need that

$$\Lambda_\alpha + E - \zeta I = \Lambda_{\alpha-\zeta}\left(I + \Lambda_{\alpha-\zeta}^{-1}E\right)$$

becomes non-invertible for some $\zeta \in \Gamma_j$. As long as $\|E\| < \epsilon$ this cannot happen, since clearly $\|\Lambda_{\alpha-\zeta}^{-1}\| = 1/\epsilon$ and thus $\|\Lambda_{\alpha-\zeta}^{-1}E\| < 1$ so that $I + \Lambda_{\alpha-\zeta}^{-1}E$ becomes invertible. This gives local Lipschitz continuity with constant 1, which easily implies global Lipschitz continuity with the same constant.

Indeed, if $B \in \mathcal{H}_n$ is another matrix we can consider the line $\{(1-t)A + tB : t \in [0,1]\}$ and use a compactness argument to extract a sequence of points $\{A_k\}_{k=0}^K$ such that $A = A_0$ and $B = A_K$ and

$$|\xi_j(B) - \xi_j(A)| = |\sum_{k=1}^{M} \xi_j(A_k) - \xi_j(A_{k-1})| \le \sum_{k=1}^{M} \|A_k - A_{k-1}\| = \|B - A\|.$$

□

Despite the fact that ξ is not even C^1, there exists a useful extension of the formula (13) which reduces the study of perturbations of some fixed eigenvalue α_j to that of a matrix which has the size of the multiplicity of α_j. In order for formulas to become easily readable, we shall henceforth denote α_j by λ and assume that following.

Definition 4.2. *The eigenvalues α to A are reordered so that $\alpha = (\lambda, \dots, \lambda, \tau_1, \dots \tau_{n-m})$, where m is the multiplicity of λ and $\tau_j \neq \lambda$ for all $j = 1, \dots, n-m$. For any matrix $E \in \mathcal{H}_n$ we furthermore introduce the block decomposition*

$$E = \begin{pmatrix} E_{11} & E_{12} \\ E_{21} & E_{22} \end{pmatrix} \tag{25}$$

where E_{11} is $m \times m$.

Given the above decomposition, we furthermore introduce a matrix B, which will play a key role in the following, by setting

$$B = E_{11} - E_{12}(\Lambda_{\tau-\lambda})^{-1}E_{21}. \tag{26}$$

To connect with the previous sections, note that if the multiplicity m of $\lambda = \alpha_j$ equals 1, then B reduces to the j'th element of $L_\xi(E) + \frac{1}{2}\mathcal{H}_\xi(E, E)$, as follows by Theorems 2.3 and 3.1. The following theorem is thus a direct extension of both theorems to the case of $m > 1$.

Theorem 4.3. *Let λ be a fixed eigenvalue of $A \in \mathcal{H}_n$ of multiplicity m, let $\{\varepsilon_j\}_{j=1}^m$ be the eigenvalus of E_{11} and $\{\beta_j\}_{j=1}^m$ those of B. Then the eigenvalues $\{\xi_j\}_{j=1}^n$ of $A + E$ can be arranged such that*

$$\xi_j = \lambda + \varepsilon_j + O\left(\|E\|^2\right) = \lambda + \beta_j + O\left(\|E\|^3\right), \quad 1 \le j \le m. \tag{27}$$

Note that $E_{11} - B = O(\|E\|^2)$ by (26), so by Theorem 4.1 it follows that β_j and ϵ_j can be ordered so that $|\epsilon_j - \beta_j| = O(\|E\|^2)$. Thus, the estimate $\xi_j = \lambda + \varepsilon_j + O\left(\|E\|^2\right)$ follows once we show that $\xi_j = \lambda + \beta_j + O\left(\|E\|^3\right)$. For this, we need a number of preparatory results. First of all, given a matrix representation

$$F = \begin{pmatrix} F_{11} & F_{12} \\ F_{21} & F_{22} \end{pmatrix} \tag{28}$$

the Schur complement of F with respect to the block F_{22} is denoted by F/F_{22} and is defined via

$$F/F_{22} = F_{11} - F_{12}F_{22}^{-1}F_{21}. \tag{29}$$

Lemma 4.4 (Schur). *The matrix F in (28) is via a change of basis similar to*

$$\begin{pmatrix} F/F_{22} & F_{12} \\ F_{22}^{-1}F_{21}(F/F_{22}) & F_{22} + F_{22}^{-1}F_{21}F_{12} \end{pmatrix}, \tag{30}$$

Proof. The matrix $J = \begin{pmatrix} I & 0 \\ F_{22}^{-1}F_{21} & I \end{pmatrix}$ is invertible with inverse $J^{-1} = \begin{pmatrix} I & 0 \\ -F_{22}^{-1}F_{21} & I \end{pmatrix}$. The result follows by computing JFJ^{-1}. □

Concerning the proof of Theorem 4.3, there is clearly no loss in generality in assuming that $\lambda = 0$ since $A + E - \lambda I$ has the same eigenvalues as $A + E$ apart from a translation by λ. Since then $A = \begin{pmatrix} 0 & 0 \\ 0 & \Lambda_\tau \end{pmatrix}$, the Schur complement of $A + E$ with respect to $\Lambda_\tau + E_{22}$ equals

$$\tilde{B} := (A + E)/(\Lambda_\tau + E_{22}) = E_{11} - E_{12}(\Lambda_\tau + E_{22})^{-1}E_{21}.$$

Recalling (26), we see that B and $\tilde{B}$ are the same apart from the appearance of E_{22} in the above inverse $(\Lambda_\tau + E_{22})^{-1}$. We will need the following result relating the eigenvalues of B with those of $\tilde{B}$. Note that both matrices are self-adjoint.

Lemma 4.5. *Let the eigenvalues of B and $\tilde{B}$, ordered non-increasingly, be denoted β and $\tilde{\beta}$ respectively. Then*

$$\beta_j = \tilde{\beta}_j + O\left(\|E\|^3\right), \quad 1 \le j \le m. \tag{31}$$

Proof. We consider the difference, recalling that $\lambda = 0$, in which case we have

$$\begin{aligned} B - \tilde{B} &= E_{12}(\Lambda_\tau + E_{22})^{-1}E_{21} - E_{12}\Lambda_\tau^{-1}E_{21} \\ &= E_{12}\Lambda_\tau^{-1}(\Lambda_\tau - (\Lambda_\tau + E_{22}))(\Lambda_\tau + E_{22})^{-1}E_{21} = -E_{12}\Lambda_\tau^{-1}E_{22}(\Lambda_\tau + E_{22})^{-1}E_{21}. \end{aligned}$$

Thus, $\|B - \tilde{B}\| = O(\|E\|^3)$ and the desired result then follows from Theorem 4.1 □

The third result we shall use in the proof of Theorem 4.3 is the Geršgorin's Circle Theorem, a version of which can be found in Ref. [8]. We recall it for completeness and remark that a recent extension to the infinite dimensional case is found in Ref. [9].

Theorem 4.6 (Geršgorin's Circle Theorem)**.** *Let M be an $n \times n$-matrix. The eigenvalues of M are contained in the union of Geršgorin discs D_j, $j = 1, \dots, n$ defined by*

$$D_j = \{z \in \mathbb{C} : |z - M(j,j)| \le R_j\}, \; R_j = \sum_{\substack{i=1 \\ i \neq j}}^{n} |M(i,j)|.$$

Furthermore if π is a permutation of $\{1, \dots, n\}$ such that $\cup_{i=1}^m D_{\pi(i)}$ is disjoint from $\cup_{i=m+1}^n D_{\pi(i)}$, then $\cup_{i=1}^m D_{\pi(i)}$ contains precisely m eigenvalues of M.

A careful inspection of the conclusion of this theorem yields the following corollary, which makes our proof of Theorem 4.3 more transparent.

Corollary 4.7. *Let $\mu_1, \dots, \mu_n$ be the eigenvalues of M ordered so that μ_j is in the union of discs containing D_j for each j. Let $\tilde{R}_j$ be the maximum of the corresponding radii (of the discs belonging to the union containing D_j). Then $|\mu_j - M(j,j)| \le 2n\tilde{R}_j$.*

Armed with these results, we can now prove Theorem 4.3. As noted after the statement, it suffices to prove the latter estimate $\xi_j = \beta_j + O\left(\|E\|^3\right)$, keeping in mind that we have set $\lambda = 0$.

Proof of Theorem 4.3. Due to Lemma 4.5, it is sufficient to prove that the eigenvalues $\left\{\xi_j\right\}_{j=1}^n$ of $A+E$ can be arranged so that

$$\xi_j = \tilde{\beta}_j + O\left(\|E\|^3\right), \quad 1 \le j \le m. \tag{32}$$

As usual, we assume that a basis has been chosen so that we can then rewrite the $A+E$ as

$$\begin{pmatrix} 0 & 0 \\ 0 & \Lambda_\tau \end{pmatrix} + \begin{pmatrix} E_{11} & E_{12} \\ E_{21} & E_{22} \end{pmatrix} = \begin{pmatrix} \tilde{B} + E_{12}(\Lambda_\tau + E_{22})^{-1}E_{21} & E_{12} \\ E_{21} & \Lambda_\tau + E_{22} \end{pmatrix}.$$

However, nothing prevents us from choosing eigenvectors in the subspace corresponding to $\lambda = 0$ so that in addition $\tilde{B}$ diagonalizes; $\tilde{B} = \Lambda_{\tilde{\beta}}$. Applying (30) from Lemma 4.4, we find that $A+E$ is similar to

$$\begin{pmatrix} \Lambda_{\tilde{\beta}} & E_{12} \\ (\Lambda_\tau + E_{22})^{-1}E_{21}\Lambda_{\tilde{\beta}} & \Lambda_\tau + E_{22} + (\Lambda_\tau + E_{22})^{-1}E_{21}E_{12} \end{pmatrix} = \begin{pmatrix} \Lambda_{\tilde{\beta}} & E_{12} \\ O(\|E\|^2) & \Lambda_\tau + O(\|E\|). \end{pmatrix}$$

For sufficiently small values of $\|E\|$ the operator $\Lambda_\tau + O(\|E\|)$ is invertible, (since Λ_τ is by Definition 4.2). Therefore, Lemma 4.4 is applicable once more, and we find that $A+E$ is similar to

$$\begin{pmatrix} \Lambda_{\tilde{\beta}} - E_{12}(\Lambda_\tau + O(\|E\|))^{-1}(\Lambda_\tau + E_{22})^{-1}E_{21}\Lambda_{\tilde{\beta}} & E_{12} \\ (\Lambda_\tau + O(\|E\|))^{-1}O(\|E\|^2)\left(\Lambda_{\tilde{\beta}} - E_{12}(\Lambda_\tau + O(\|E\|))^{-1}O(\|E\|^2)\right) & \Lambda_\tau + O(\|E\|) \end{pmatrix}$$
$$= \begin{pmatrix} \Lambda_{\tilde{\beta}} + O(\|E\|^3) & E_{12} \\ O(\|E\|^3) & \Lambda_\tau + O(\|E\|) \end{pmatrix} \tag{33}$$

where we used that $\tilde{\beta} = O(\|E\|)$. We now apply Theorem 4.6 to the final matrix. Clearly, since τ has nonzero elements and $\tilde{\beta}$ is near 0, for small $\|E\|$, it is possible to choose some $\delta > 0$ such that the Geršgorin discs D_j for $j \le m$ are disjoint from the corresponding discs with $j > m$. Thus, Corollary 4.7 implies that $\left(\xi_j\right)_{j=1}^n$ can be ordered so that $|\xi_j - \tilde{\beta}_j| = O\left(\|E\|^3\right)$ (since the off-diagonal elements of the first m columns of the above matrix are $O\left(\|E\|^3\right)$). This establishes (32) and thus, the proof is complete. □

We remark that the estimate in Theorem 4.3 can be made slightly more precise as follows

$$|\xi_j - \lambda + \beta_j| \le 5m\delta^{-2}\|E\|^3 + O\left(\|E\|^4\right), \quad 1 \le j \le m, \tag{34}$$

where δ as above is the isolation distance of λ. This is shown in Ref. [10], where a version of Theorem 4.3 for operators on separable Hilbert spaces is found as well. This publication also includes an extension to perturbations of the singular value decomposition.

We now consider an example. We take the same α as we used in **Figures 2–4**, except that at the problematic positions $J = (9,10,15,16,17,18)$ we average so that $\alpha(j) = \alpha(j+1)$ for $j = 9,15,17$. The estimates for $\xi_j(E)$ from Theorems 2.3 and 3.1 now become completely unreliable at the problematic positions, *but still work well for the remaining j's*. In **Figure 5**, we therefore computed approximations of $\xi_j(E)$ using the "old" expressions for $j \notin J$. However, we use the new expression (27) from Theorem 4.3 to approximate the problematic indices $j \in J$.

As is clear to see from the middle and right plots in **Figure 5**, there is a tremendous improvement in accuracy. More precisely, comparing with the middle plot of **Figure 2** and the left of **Figure 4**, respectively, we see that there is roughly a 20-fold improvement in the first-order estimates, and a 500-fold improvement for the second-order estimates. This indicates that the comparatively poor performance in the previous plots was caused principally by the estimates related to the problematic positions J, indicating that one needs to be careful when α has eigenvalues which take almost the same value. To further underscore this observation, we include in **Figure 6** a plot showing the average of the modulus of the error for each index j, using the same setup as in **Figure 5**. As is plain to see, errors are of the order 10^{-11} for the first-order formulas, and 10^{-15} for the second-order formulas, independent of whether we are at a degenerate eigenvalue of A (i.e., $j \in J$) or not.

As a final remark we note that there is no hope of a similar theorem for the eigenvectors. Indeed, a reasonable conjecture in the light of Theorem 4.3 would be that also the eigenvectors of B carry some information about corresponding eigenvectors for $A + E$, but this is not the case. Indeed, consider

$$A + E = \begin{pmatrix} 0 & 0 & 0 \\ 0 & 0 & 0 \\ 0 & 0 & 1 \end{pmatrix} + \begin{pmatrix} e^2 & ef & e \\ ef & f^2 & f \\ e & f & 0 \end{pmatrix}$$

In this case, $B = 0$ and $(f, -e, 0)^T$ is an eigenvector of $A + E$. However, pick any matrix $F \in \mathcal{H}_n$ which is zero on the last row and column, and consider

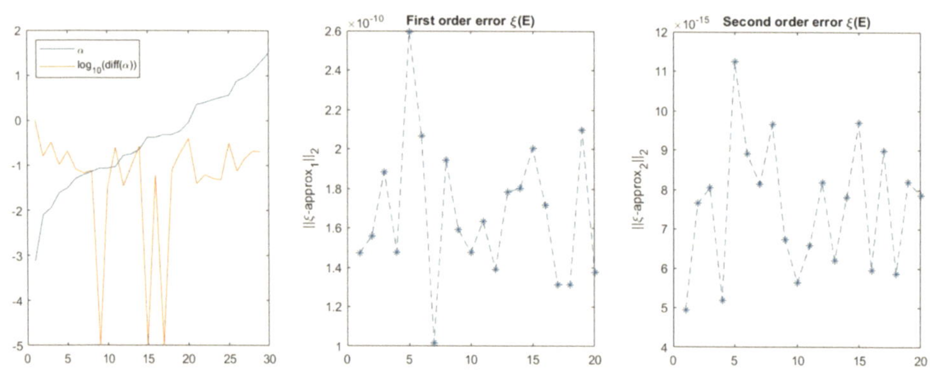

Figure 5.
*Illustration of Theorem 4.3. Left figure displays the eigenvalues α used in this example, along with a $\log_{10}$-plot of the difference of adjacent eigenvalues (where the value is $-\infty$ for positions 9, 15, and 17). The middle graph shows the error of the first-order estimate in formula (27), for the same 20 realizations of perturbations E with $\|E\| = 10^{-5}$ that we used in **Figure** 2, and the right graph shows the corresponding second-order error. For further details of how the approximations were computed, see the main text.*

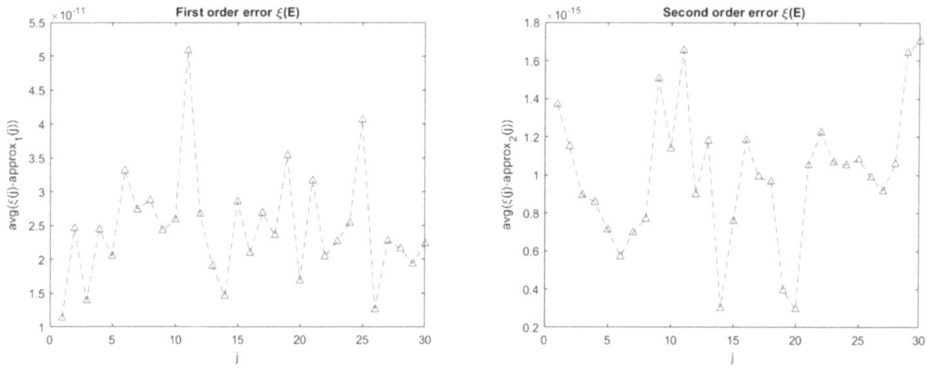

Figure 6.
The average of the pointwise estimates, using Theorems 2.3 and 3.1 for the simple eigenvalues and Theorem 4.3 for the degenerate eigenvalues $j \in J$.

$t = \mapsto A + E + tF$. Since the eigenvectors are analytic in t, due to Rellich's theorem, we can pick a very small t so that there exists an eigenvector arbitrarily close to $(f, -e, 0)^T$. However, in this case the matrix B, which equals the upper left 2x2 corner of tF, can have any given eigenvectors. In a similar fashion, one can prove that the eigenvectors of E_{11} may be totally misleading.

5. Asymptotic behavior of analytic perturbations

In stark contrast to the situation described above, the eigenvectors become real analytic if we restrict attention to perturbations of A of the form $A + tE$ where $t \in \mathbb{R}$ is variable and $E \in \mathcal{H}_n$ is fixed. In the light of this and Theorem 2.3, one could hope that the first-order coefficients in the expansion of U would still be given by $M \diamond E$ for some suitable modification of M. Surprisingly, this turns out to be false, and since we have not been able to locate the expression for the correct modification in any of the key references on perturbation theory, we include a section with explicit formulas for this case.

More generally, in some applications it is of interest to consider analytic perturbations and study the asymptotic behavior of the eigenvalues (and vectors) as the parameter goes to 0 (see [11] and the references therein). We shall thus consider a real analytic function $A(t) = A + \sum_{k=1}^{\infty} E_{(k)} t^k$ and let $\xi(t)$ be the real analytic eigenvalues and $V(t)$ the corresponding eigenvectors (which exist according to Rellich's theorem, i.e., the ordering can be chosen so that both become real analytic). As before we can assume that a basis has been chosen so that $A = \Lambda_\alpha$ for some vector α. The trick with making an "ansatz" and then back out the coefficients of interest, as we did in Theorems 2.3–3.2, can be perfectly adapted to this framework.

Let $\xi(t) = \alpha + \sum_{k=1}^{\infty} \xi_{(k)} t^k$ and $V(t) = I + \sum_{k=1}^{\infty} V_{(k)} t^k$, and assume for simplicity that each $V_{(k)}$ vanish on the diagonal, so that $\texttt{diag}(V(t)) = [1, \ldots, 1]^T$. We remark that the notation is not entirely optimal here, but since we have already used, for example, ξ_j to denote the j'th component of the vector $\xi(t)$ in $\mathbb{R}^n$, we cannot use the same notation for the coefficients in the series expansion. We therefore opted for $\xi_{(k)}$ where the parenthesis (hopefully) helps to avoid confusion.

Consider the equation

$$\left(I+\sum_{k=1}^{\infty}V_{(k)}t^k\right)\left(\Lambda_\alpha+\sum_{k=1}^{\infty}\Lambda_{\xi_{(k)}}t^k\right)=\left(\Lambda_\alpha+\sum_{k=1}^{\infty}E_{(k)}t^k\right)\left(I+\sum_{k=1}^{\infty}V_{(k)}t^k\right) \tag{35}$$

Equating the first-order terms gives $V_{(1)}\Lambda_\alpha+\Lambda_{\xi_{(1)}}=E_{(1)}+\Lambda_\alpha V_{(1)}$ and thus

$$\left[V_{(1)},\Lambda_\alpha\right]+\Lambda_{\xi_{(1)}}=E_{(1)}, \tag{36}$$

which is just Eq. (8) in the new setting. However, when some eigenvalue α_k is degenerate, the commutator vanishes on the entire block of indices $(i,j)\in I_{\alpha_k}\times I_{\alpha_k}$, where $I_{\alpha_k}=\{j:=\alpha_j=\alpha_k\}$. In this case, the equation is inconclusive, that is, $V_1(i,j)$ can take any value for $i\neq j\in I_{\alpha_k}$, and still satisfy (36). To determine these values, we need to look at higher-order identities. If the eigenvalues of the submatrix $\left(E_{(1)}(i,j)\right)_{i,j\in I_{\alpha_k}}$ are distinct, then it suffices to look at the second-order term of (35). We satisfy with stating explicit formulas for this particular case and leave the extension to more general cases (which can be obtained using the recursive method described here) to the interested reader.

Theorem 5.1. *Let $A(t)$, $\xi(t)$ and $V(t)$ be as above, and assume that the basis has been chosen so that $A(0)=\Lambda_\alpha$ and moreover so that the submatrix $\left(E_{(1)}(i,j)\right)_{i,j\in I_{\alpha_k}}$ is diagonal for each eigenvalue α_k. Then the eigenvalues $\left(\xi_j(t)\right)_{j=1}^n$ can be ordered so that*

$$\xi_j(t)=\alpha_j+tE_{(1)}(j,j)+t^2\left(E_{(2)}(j,j)+\sum_{l\notin I_{\alpha_j}}\frac{\left|E_{(1)}(j,l)\right|^2}{\alpha_j-\alpha_l}\right)+O(t^3) \tag{37}$$

whereas $V(t)=I+tV_{(1)}+O(t^2)$ where

$$V_{(1)}(i,j)=\begin{cases}E_{(1)}(i,j)/(\alpha_j-\alpha_i), & \alpha_i\neq\alpha_j,\\ 0, & i=j,\\ \left(E_{(2)}(i,j)+\sum_{l:\alpha_l\neq\alpha_j}\frac{\overline{E_{(1)}(l,i)}E_{(1)}(l,j)}{\alpha_j-\alpha_l}\right)/\left(E_{(1)}(i,i)-E_{(1)}(j,j)\right), & \textit{else}.\end{cases} \tag{38}$$

Proof. Just as in the proofs of Theorem 2.3 and 3.1, the linear term $E_{(1)}(j,j)$ in (37) can be read immediately from the Eq. (36), as well as the values $E_{(1)}(i,j)\left(\alpha_j-\alpha_i\right)^{-1}$ in the definition of $V_{(1)}$ for indices such that $\alpha_i\neq\alpha_j$. Also, the fact that $V_{(1)}$ should equal 0 on the diagonal follows from the assumption $\text{diag}(V(t))=[1,\ldots,1]^T$. It remains to prove the second-order term in (37) as well as the bottom line of (38). Both of these can be extracted by considering the second-order term of (35), which reads

$$V_{(2)}\Lambda_\alpha+V_{(2)}\Lambda_{\xi_{(1)}}+\Lambda_{\xi_{(2)}}=E_{(2)}+E_{(1)}V_{(1)}+\Lambda_\alpha V_{(2)}. \tag{39}$$

By what we have already shown we have $\xi_{(1)}=\texttt{diag}(E_{(1)})$ and, introducing $E_{(1)}^{od}$ for the matrix containing the off-diagonal elements of $E_{(1)}$ (i.e., $E_{(1)}-\Lambda_{\texttt{diag}(E_{(1)})}$), we can write $E_{(1)}=\Lambda_{\texttt{diag}(E_{(1)})}+E_{(1)}^{od}$ and thus (39) implies

$$\left[V_{(2)}, \Lambda_\alpha\right] + \left[V_{(1)}, \Lambda_{\mathtt{diag}E_{(1)}}\right] + \Lambda_{\xi_{(2)}} = E_{(2)} + E_{(1)}^{od} V_{(1)}. \tag{40}$$

We recall that $E_{(1)}^{od}$ is zero (by our choice of basis) for all indices (l,j) within the same "block" $I_{\alpha_j} \times I_{\alpha_j} = \left\{(l,j) : \alpha_l = \alpha_j\right\}$. Outside of these blocks, we know (by what we already established) that $V_{(1)}(l,j) = \left(\alpha_j - \alpha_l\right)^{-1} E_{(1)}(l,j)$, and thus, for (i,j) inside of the same block we have

$$\left(E_{(1)}^{od} V_{(1)}\right)(i,j) = \sum_{l:\alpha_l \neq \alpha_j} E_{(1)}(i,l)\left(\alpha_j - \alpha_l\right)^{-1} E_{(1)}(l,j).$$

Thus, since the commutator $\left[V_{(2)}, \Lambda_\alpha\right]$ vanishes for (i,j) in the same block, eq. (40) entails that $\left[V_{(1)}, \Lambda_{\mathtt{diag}E_{(1)}}\right] + \Lambda_{\xi_{(2)}} = E_{(2)} + E_{(1)}^{od} V_{(1)}$ or, written out, that

$$\left(E_{(1)}(j,j) - E_{(1)}(i,i)\right) V_{(1)}(i,j) + \xi_{(2)}(i)\delta_{i,j} = E_{(2)}(i,j) + \sum_{l:\alpha_l \neq \alpha_j} E_{(1)}(i,l)\left(\alpha_j - \alpha_l\right)^{-1} E_{(1)}(l,j),$$

where δ_{ij} denotes the "Kronecker delta-symbol." The second-order term in (37) as well as the bottom row of (38) follow immediately from this. □

We illustrate with a 4x4 numerical example in which $\alpha = (2,1,1,1)$, whereas $E_{(1)}$ and $E_{(2)}$ were randomly generated in $\mathcal{H}_n$. We tested the formula on an interval $t \in [0,0.1]$ see **Figure 7**. As is plain to see, the graphs are rather uninteresting, almost linear, but if one goes beyond $t = 0.1$ the behavior quickly becomes very nonlinear, indicating that formula (37) should be used carefully for larger perturbations. To fully appreciate the effect of the second-order term, we display also the difference between $\xi(t)$ and its approximation with two and three terms, respectively, from which it follows that we gain a factor of about 10 in accuracy in this particular example. This is, however, rather case specific.

Figure 8 comes from the same example, but instead shows accuracy of the off-diagonal elements in (38). On the full interval $[0,0.1]$ the approximations quickly deteriorate, indicating numerically that estimating the eigenvectors is more unstable

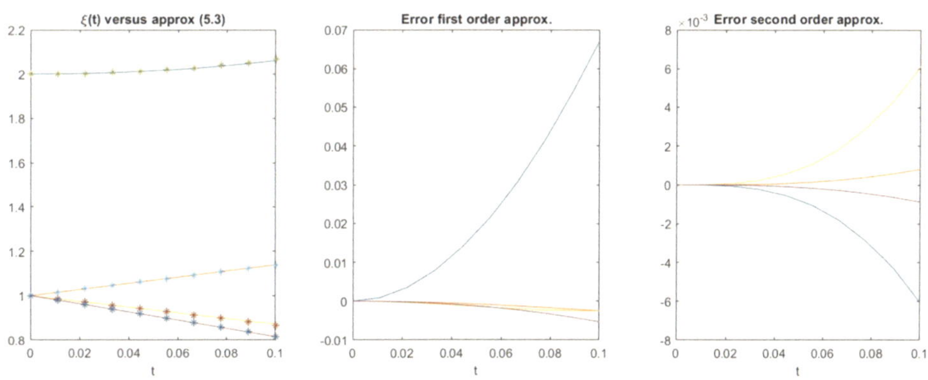

Figure 7.
*We approximated ξ by formula 37 (omitting the ordo-term) and display the result to the left (where approximation values are marked by *). In the middle graph, we see the difference between $\xi(t)$ and the approximation using only the constant and linear term from (37), and to the right we used all three terms.*

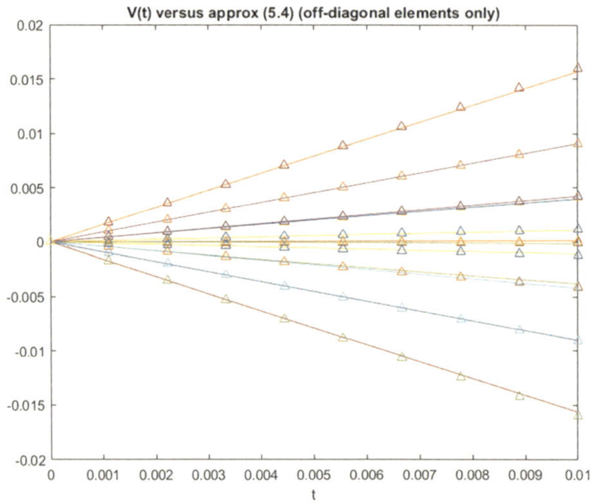

Figure 8.
Difference between $V(t)$ and its linear approximation according to (38), where we only display the off-diagonal terms (since the diagonal is constant).

than estimating eigenvalues. We therefore we only display the smaller interval $[0,0.01]$, which is enough to see that the formulas are correct.

6. Historical remarks

The coefficients in the expansion of $\xi(t)$ (37), in the case $E_{(2)} = 0$, are the famous Rayleigh-Schrödinger coefficients, which date back at least to the 1870's and the book "The theory of sound" by Lord Rayleigh [1]. However, despite appearing in many classical works, we have not been able to locate anywhere in the modern literature the corresponding formula (38) for $U(t)$. Both (37) and (38) (for the line-perturbation $A + tE$), also in the case of degenerate eigenvalues, do appear in the classic by Courant and Hilbert [7], for the particular case when A is a self-adjoint differential equation on an infinite-dimensional space, in the context of analyzing vibrations (see Ch. 5.13). It is also from here the author of the present text got the idea of backing out the coefficients by a simple ansatz and then successively solving for the terms, relying on Rellich for convergence. However, the setting in Ref. [7] is so particular and the way it is presented rather hard to decipher, that I suspect most readers who glance through looking for general results on matrix perturbation theory probably miss that what is presented is a general recipe for computing derivatives of any order of both $\xi(t)$ and $U(t)$. Henceforth, I thought it was worthwhile to write this down in a concise and modern manner.

Modern influential textbooks on the topic, such as Reed and Simon [6], use integral formulas in the complex plane, rather than the simpler ansatz that underlies the proofs in the preceding sections, and it is not easy to derive, for example, (38) from these. It is noteworthy, however, that Reed and Simon plow through to compute even the fourth-order term in the expansion of $\xi(t)$ (albeit for the case of a line-perturbation $A + tE$, assuming distinct eigenvalues of A). The other major reference on the subject by Kato [12], on the other hand, skips explicit formulas altogether. One may suspect that the method using the ansatz was well known in the early twentieth century and later got pushed away by more elegant modern techniques relying on complex integration of resolvents, but this is only speculation. In either case, the fact that the eigenvalues and vectors are analytic as a function of *all* coefficients (Theorem 2.1 above) can be found, for example, in Theorem 5.16 of Kato's bible on the topic, but he does not give any formulas for the Frechét derivatives. These formulas, as presented in Theorems 2.3–3.2, have not made its way to any textbook on matrix theory that we are aware of, see Refs. [2, 8, 13–16], and we therefore thought it would be useful for the community to have a reference for these facts. Again, it is possible that these results were known to the "masters of old," given that the proofs presented here are fairly elementary and build on the ideas of Courant, Hilbert and Rellich.

In any case, the extention to the degenerate case (Theorem 4.1) is a recent contribution. The result was announced in Ref. [17] (along with the formula (38)) and published in [10] (in collaboration with O. Rubin) with an improved proof that extends to the case of self-adjoint operators on separable Hilbert spaces. The proof presented here is a simplified version of the proof found in [10], adapted to the matrix case. Both works [10, 17] contain more historical notes and further references to the research literature on the topic.

Author details

Marcus Carlsson
Centre for Mathematical Sciences, Lund University, Lund, Sweden

*Address all correspondence to: marcus.carlsson@math.lu.se

References

[1] Rayleigh L. The Theory of Sound. Vol. 1. London: Macmillan; 1877

[2] Baumgärtel H. Analytic Perturbation Theory for Matrices and Operators. Basel: Birkhäuser Verlag; 1985

[3] Kato T. Perturbation Theory for Linear Operators. Vol. 132. Berlin Heidelberg: Springer Science & Business Media; 2013

[4] Rellich F. Störungstheorie der spektralzerlegung. Mathematische Annalen. 1937;**113**(1):600-619

[5] Rellich F, Berkowitz J. Perturbation Theory of Eigenvalue Problems. Boca Raton: CRC Press; 1969

[6] Reed M, Simon B. Methods of Modern Mathematical Physics IV: Analysis of Operators. London: Academic Press; 1978

[7] Courant R, Hilbert D. Methods of Mathematical Physics. CUP Archive; 1955

[8] Horn R, A, Johnson CR. Matrix Analysis. Cambridge: Cambridge University Press; 1990

[9] Carlsson M, Rubin O. A Hilbert-space variant of geršgorin's circle theorem. Mathematische Nachrichten. 2024:3095-3106

[10] Carlsson M, Rubin O. On perturbation of operators and Rayleigh-Schrödinger coefficients. Complex Analysis and Operator Theory. 2024; **18**(3):47

[11] Barthelmé S, Usevich K. Spectral properties of kernel matrices in the flat limit. SIAM Journal on Matrix Analysis and Applications. 2021;**42**(1):17-57

[12] Kato T. Upper and lower bounds of eigenvalues. Physical Review. 1950; **77**(3):413

[13] Bhatia R. Matrix Analysis. Vol. 169. Heidelberg, New York: Springer Science & Business Media; 2013

[14] Parlett B, N. The Symmetric Eigenvalue Problem. Vol. 20. Philadelphia: SIAM; 1998

[15] Stewart G, W, Sun J-g. Matrix Perturbation Theory. New York: Elsevier Science; 1990

[16] Wilkinson JH. The Algebraic Eigenvalue Problem. Vol. 87. Oxford: Clarendon Press; 1965

[17] Carlsson M. Perturbation theory for the spectral decomposition of hermitian matrices, arXiv 1809.09480, 2018

Chapter 5

Hybrid Parallel Scheme for Eigenvalue Problems Using Multiplicative Calculus

Mudassir Shams and Bruno Carpentieri

Abstract

The complexity of the nonlinear eigenvalue problem is too great for traditional analytical methods, particularly in engineering applications where dynamic behaviors and material nonlinearities are crucial. For this reason, this chapter presents a novel hybrid parallel technique based on multiplicative calculus as an effective way to handle scale-invariant processes without dealing with sum divergence issues. It offers improved precision and computational stability, making it especially appropriate for complex systems where ratios—rather than differences—are crucial. The results are verified in the nonlinear dynamical analysis of frame structures, a highly relevant applied engineering problem involving substantial deformations and material nonlinearities. The proposed formulation achieves a convergence order of three. The numerical findings reveal that our method performs better than existing strategies in the literature, with considerable improvements in residual error, computational efficiency, stability, and CPU time. These findings show the method's potential application in handling real-world engineering challenges such as dynamic structural analysis.

Keywords: multiplicative calculus, parallel scheme, error graph, computational efficiency, Eigenvalue problem

1. Introduction

Eigenvalue issues have become crucial tools for analysis and design since they are used in the majority of applications that include science and engineering [1, 2]. In structural engineering, the natural frequencies of vibrating systems are properly matched to prevent the disaster of resonance, which is caused by the potential failure of structures. Schrödinger equation [3] in quantum mechanics presents eigenvalue issues, in which the eigenvalues stand in for physical entities like the energy levels of atoms and molecules. Eigenvalues are also essential for stability analysis in fluid dynamics and control systems [4], where solutions must either diverge or remain stable [4]. Electrical engineers encounter eigenvalue problems while solving circuits and systems of linear differential equations [5]. Chemists apply eigenvalues to approximate the vibrations of molecules, which basically acts as a frame of reference

 IntechOpen

in interpreting spectroscopic data [6]. Likewise, for an engineer, eigenvalues are found in the determination of the stiffness and flexibility properties of a material to design better, stronger, and efficient structures [7]. Generally, eigenvalue problems are extremely helpful in modeling and optimization, and even in prediction of complicated behavior in any system in multiple disciplines [8–10].

In structural dynamics [11], which corresponds to the study of the analyses of nonlinear frame structures such as buildings, bridges, and mechanical systems, it becomes very important to know how these systems work in response to dynamic loads resulting from wind, earthquake, and moving loads. The next nonlinear problem describes how these above kinds of structures work dynamically as:

$$\begin{cases} E_1^{[*]}I_1^{[*]}\left(\frac{\partial^6 U(x,t)}{\partial x^6}+\frac{\partial^4 U(x,t)}{\partial x^4}\right)+\rho^{[*]}A^{[*]}\frac{\partial^2 U(x,t)}{\partial t^2}+\alpha^{[*]}U(x,t) \\ +h_{[n]}(U(x,t))=0, \end{cases} \tag{1}$$

where $h_{[n]}(U(x,t))$ is a nonlinear functions,

$E_1^{[*]}I_1^{[*]}$ represents the rigidity or flexibility of the frame structure,

$\rho^{[*]}$ is the density of the frame structure,

$A^{[*]}$ is the cross-sectional area, and

$\alpha^{[*]}$ is the stiffness constant.

We analyze the stability and behavior of the frame structure by solving the governing equations of motion, which usually are nonlinear due to large deformations, material nonlinearity, or geometric effects. The systems can be modeled using differential equations. In the case of the natural frequencies and vibrational modes, this always leads to an eigenvalue problem. Dynamic analysis of nonlinear frame structures is important as these present how complex systems like buildings and bridges respond to dynamic loads like wind and earthquake. The nonlinearities of the systems arise from either a large deformation or the features of the material, which creates complications in solving the governing equations with the traditional exact or analytical methods [12, 13]. Numerical methods offer this flexibility to model complex boundary conditions in coupled systems to be solved efficiently. These algorithms are particularly good at solving nonlinear eigenvalue problems that cannot be solved analytically.

To reduce the complexity of the frame structure problem, we apply transformations such as the Fourier transform [13]. This transforms the problem into a system of ordinary differential equations. Utilizing eigenvalues, we reformulated these problems in the form of nonlinear equations:

$$h(x)=0, \tag{2}$$

Often of degree five or higher, making it impractical to solve the eigenvalue problem analytically, as there is no general algebraic solution for polynomials of degree greater than five [14]. In such cases, parallel algorithms are necessary to approximate the eigenvalues and eigenvectors. Consequently, we employ a parallel technique that simultaneously approximates all eigenvalues and their corresponding eigenfunctions. To further enhance the efficiency and stability of the parallel scheme [15], hybrid parallel techniques [16, 17] are utilized, proving highly effective in solving Eq. (1).

Multiplicative calculus [18], developed in the twentieth century, extends classical calculus by focusing on growth rates expressed in multiplicative terms rather than differences. The primary aim of this chapter is to develop an efficient parallel scheme based on multiplicative calculus for simultaneously determining all eigenvalues and correlation functions. This approach uses multiplicative derivatives and integrals to provide a more accurate and concise description of real-world phenomena where ratios and proportionality are more relevant than differences. Multiplicative calculus provides particular frameworks for numerical methods to solve nonlinear equations, as opposed to the traditional, so-called additive calculus, which defines differentiation and integration. Traditional calculus uses sum and difference, while multiplicative calculus considers product and quotient, enabling you to analyze with other tools some functional classes growth and decay processes. This distinction can enable methods that are better appropriate for specific types of nonlinear equations, particularly those with exponential or geometric form.

1.1 Multiplicative calculus methods: Strengths

- *Consistency with actual processes*: Since multiplicative models do represent possibilities of growth in nature and in finance, they are likely to be more appropriate in situations in which additive changes are simply not possible.

- *Fewer iterations:* In scenarios in which the behavior is exponentially modeled, multiplicative methods will have to carry out many fewer iterations since they do not warp nonlinear, multiplicative growth into a linear perspective.

- *Better numerical stability:* Errors are treated in proportion to their values. This actually helps limit the propagation of errors and is an important property for such sensitive calculations which can iterate many times.

- *Convergence and error behavior:* Further to the conventional additive effect—for example, multiplicative methods analyze errors multiplicatively, which is sometimes expressed as a ratio or percentage. Stability and convergence could be enhanced by this way of view for issues where errors increase after a number of repetitions. For example, if the root of a function is exponential, such a multiplicative technique will act in accordance with the function, reducing error propagation and potentially increasing convergence speed.

- *Implementations in diverse technical applications:* Traditional techniques assume that functions may be well approximated using linear (additive) increments. While multiplicative calculus-based numerical approaches are preferable for the nonlinear problems where proportional or exponential increases are dominating, such as population expansion, financial growth, nonlinear biomedical engineering problems, eigenvalue problems, and compound interest models. In these situations, they can usually converge faster and with more accuracy.

Definition: A function $\wp : \varpi \subset \mathbb{R} \to \mathbb{R}$ is said to be multiplicatively differentiable if its multiplicative derivative exists, defined as:

$$\wp^{[*]}(x) = \frac{d^{[*]}\wp}{dx} = \lim_{t \to 0} \left(\frac{\wp(x+t)}{\wp(x)} \right)^{\frac{1}{t}}, \tag{3}$$

where $t > 0$ and the derivative of $\wp$ at x exists. The n-th multiplicative derivative is then defined as [19]:

$$\wp^{[*]}(x) = e^{(\ln\wp)'(x)} \tag{4}$$

where $\ln\circ\wp = \ln(\wp(x))$. Higher-order multiplicative derivatives are similarly defined. For instance, the second-order multiplicative derivative is:

$$\wp^{[**]}(x) = e^{(\ln\circ\wp)''}(x) \tag{5}$$

and in the general case:

$$\wp^{[*](n)}(x) = e^{(\ln\circ\wp)^{(n)}}(x), n = 0, 1, \ldots \tag{6}$$

where $n = 0$, no multiplicative derivative exists there and represents the original function $\wp(x) = 1$.

Definition: Let $\wp : \varpi \subset \mathbb{R} \to \mathbb{R}^{+}$ be a positive, nonlinear function. The multiplicative nonlinear equation [20] is defined as:

$$\wp(x) = 1. \tag{7}$$

Some key properties of multiplicative differentiation for multiplicatively differentiable functions q and $\wp$ are as follows:

$$\begin{aligned}
(C)^{[*]} &= 1,\\
(C\wp)^{[*]}(x) &= Cq^{[*]}(x),\\
(\wp\circ q)^{[*]}(x) &= \wp^{[*]}(x)q^{[*]}(x),\\
\left(\frac{\wp}{q}\right)^{[*]}(x) &= \frac{\wp^{[*]}(x)}{q^{[*]}(x)},\\
(\wp^{\psi})^{[*]}(x) &= q^{[*]}(x)^{\psi(x)}\wp(x)^{\psi'(x)},\\
(\wp\circ\psi)^{[*]}(x) &= \wp^{[*]}\psi(x)^{\psi'(x)}.
\end{aligned}$$

The multiplicative Taylor theorem [21] plays a crucial role in constructing new numerical schemes for solving the nonlinear problem (2).

Theorem 1: Let $\wp : \varpi \to \mathbb{R}$ be a function that is multiplicatively differentiable $(n+1)$ times over an open interval ϖ. Then for any $x, x + a \in \varpi$, there exists a number $\eta \in (0, 1)$ such that:

$$\wp(x + a) = \prod_{r=0}^{n}\left(\wp^{[*](r)}(x)\right)^{\frac{a^r}{r!}}\left(\wp^{[*](n+1)}(x + \eta a)\right)^{\frac{a^{n+1}}{(n+1)!}}. \tag{8}$$

This theorem is used to ensure the convergence of the parallel scheme.

2. Multiplicative calculus-based parallel scheme

In multiplicative calculus, numerical methods simplify the handling of scale-invariant processes, which can pose challenges for classical calculus due to complex transformations. Additionally, multiplicative approaches avoid problems such as sum divergence, enhancing efficiency in systems with large-scale variability. These methods also improve the precision of error estimation in systems where ratios, rather than absolute values, are more relevant. Multiplicative interactions are particularly important in fields such as biology, finance, fractal analysis, and differential equations. Furthermore, Singh et al. [22] proposed a multiplicative version of the Schröder method, defined as:

$$u^{[r]} = x^{[r]} - \frac{\ln(\wp(x^{[r]}))\ln(\wp^{[*]}(x^{[r]}))}{(\ln(\wp^{[*]}(x^{[r]})))^2 - \ln(\wp^{[**]}(x^{[r]}))\ln(\wp(x^{[r]}))}., \tag{9}$$

which has a convergence order of 2.

Robust computational procedures, known as parallel computing approaches, are employed to simultaneously find all possible solutions to nonlinear equations. The uniqueness of these methods lies in their simplicity: They iteratively improve approximations, regardless of how close or far they are from the exact solutions, using computer multiprocessing to compute all solutions in parallel. Among these, the Weierstrass method is one of the best derivative-free parallel algorithms. It is described as:

$$x_i^{[r+1]} = x_i^{[r]} - \Delta\left(x_i^{[r]}\right), \tag{10}$$

where

$$\Delta\left(x_i^{[r]}\right) = \frac{h\left(x_i^{[r]}\right)}{\prod\limits_{\substack{j=1 \\ j\neq i}}^{n}\left(x_i^{[r]} - x_j^{[r]}\right)}, \qquad (i,j = 1, \dots, n), \tag{11}$$

is known as Weierstrass' correction. This method exhibits local quadratic convergence.

In 1977, Ehrlich [23] introduced the following third-order simultaneous method:

$$x_i^{[r+1]} = x_i^{[r]} - \frac{1}{\frac{1}{N_i\left(x_i^{[r]}\right)} - \sum\limits_{\substack{j=1 \\ j\neq i}}^{n}\left(\frac{1}{\left(x_i^{[r]} - x_j^{[r]}\right)}\right)}, \tag{12}$$

where $x_j^{[r]} = u_j^{[r]}$ is used as a correction in (4).

Next, consider the well-known single-root finding scheme presented by Weerakoon et al. [24], which is given as:

$$v^{[r]} = x^{[r]} - \left[\frac{2h\left(x^{[r]}\right)}{h'(x^{[r]}) + h'(y^{[r]})}\right], \tag{13}$$

where $y^{[r]} = x^{[r]} - \frac{h\left(x^{[r]}\right)}{h'(x^{[r]})}$. Taking the natural logarithm of (11) and differentiating, we obtain:

$$\frac{\Delta'(x)}{\Delta(x)} = \frac{h'\left(x^{[r]}\right)}{h\left(x^{[r]}\right)} - \sum_{\substack{j=1 \\ j \neq i}}^{n} \left(\frac{1}{x_i^{[r]} - x_j^{[r]}} \right), \tag{14}$$

or equivalently:

$$\frac{\Delta(x)}{\Delta'(x)} = \frac{1}{\frac{h'\left(x^{[r]}\right)}{h\left(x^{[r]}\right)} - \sum_{\substack{j=1 \\ j \neq i}}^{n} \left(\frac{1}{\left(x_i^{[r]} - x_j^{[r]}\right)} \right)}, \tag{15}$$

Using (15), (11), and (9) as corrections in (13), we develop a new parallel scheme for simultaneously finding all eigenvalues of (1) as:

$$v_j^{[r]} = x_j^{[r]} - \frac{\left(\frac{2\prod_{\substack{j=1 \\ j \neq i}}^{n} \left(x_i^{[r]} - x_j^{[r]}\right)}{\prod_{\substack{j=1 \\ j \neq i}}^{n} \left(x_i^{[r]} - x_j^{[r]}\right) - \prod_{\substack{j=1 \\ j \neq i}}^{n} \left(y_i^{[r]} - y_j^{[r]}\right)} \right)}{\left[\frac{h'\left(x_i^{[r]}\right)}{h\left(x_i^{[r]}\right)} - \sum_{\substack{j=1 \\ j \neq i}}^{n} \left(\frac{1}{\left(x_i^{[r]} - u_j^{[r]}\right)} \right) \right]}, \tag{16}$$

where $y_i^{[r]} = x_i^{[r]} - \frac{1}{\frac{h'\left(x_i^{[r]}\right)}{h\left(x_i^{[r]}\right)} - \sum_{\substack{j=1 \\ j \neq i}}^{n} \left(\frac{1}{\left(x_i^{[r]} - u_j^{[r]}\right)} \right)}$ and

$$u_j^{[r]} = x_j^{(r)} - \frac{\ln\left(\wp\left(x_j^{(r)}\right)\right) \ln\left(\wp^{[*]}\left(x_j^{(r)}\right)\right)}{\left(\ln\left(\wp^{[*]}\left(x_j^{(r)}\right)\right)\right)^2 - \ln\left(\wp^{[**]}\left(x_j^{(r)}\right)\right) \ln\left(\wp\left(x_j^{(r)}\right)\right)}..$$

The following theorem defines the local order of convergence for the inverse fractional scheme.

Theorem 2: Let $\zeta_1, \ldots, \zeta_\sigma$ be simple roots of a nonlinear equation, and assume that the initial distinct estimates $x_1^{[0]}, \ldots, x_n^{[0]}$ are sufficiently close to the true roots. Then, the $\text{ZM}^{[*]}$ method achieves a convergence order of three.

Proof: Let $\epsilon_i = x_i^{[r]} - \zeta_i, \epsilon_y = y_i^{[r]} - \zeta_i$, and $\epsilon_i^{[*]} = v_i^{[r]} - \zeta_i$ represent the errors in $x_i^{[r]}, y_i^{[r]}$, and $v_i^{[r]}$, respectively. From the first-step of the $\text{ZM}^{[*]}$ method, we have:

$$y_i^{[r]} - \zeta_i = x_i^{[r]} - \zeta_i - \frac{1}{\frac{h'\left(x_i^{[r]}\right)}{h\left(x_i^{[r]}\right)} - \sum_{\substack{j=1 \\ j \neq i}}^{n} \left(\frac{1}{\left(x_i^{[r]} - u_j^{[r]}\right)} \right)}, \tag{17}$$

$$\epsilon_y = \epsilon_i - \frac{1}{\frac{1}{x_i^{[r]} - \zeta_i} + \sum_{\substack{j=1 \\ j \neq i}}^{n} \left(\frac{1}{\left(x_i^{[r]} - \zeta_j\right)} \right) - \sum_{\substack{j=1 \\ j \neq i}}^{n} \left(\frac{1}{\left(x_i^{[r]} - u_j^{[r]}\right)} \right)}, \tag{18}$$

$$\epsilon_y = \epsilon_i - \frac{1}{\frac{1}{\epsilon_i} + \sum_{\substack{j=1 \\ j \neq i}}^{n} \left(\frac{-u_j^{[r]} + \zeta_j}{\left(x_i^{[r]} - \zeta_j\right)\left(x_i^{[r]} - u_j^{[r]}\right)} \right)}, \tag{19}$$

$$\epsilon_y = \epsilon_i - \frac{\epsilon_i}{1 + \epsilon_i \sum_{\substack{j=1 \\ j \neq i}}^{n} \left(\frac{-\epsilon_j}{\left(x_i^{[r]} - \zeta_j\right)\left(x_i^{[r]} - u_j^{[r]}\right)} \right)}, \tag{20}$$

$$\epsilon_y = \epsilon_i - \frac{\epsilon_i}{1 + \epsilon_i \sum_{\substack{j=1 \\ j \neq i}}^{n} \epsilon_j^2 Q_{ij}^{[*]}}, \tag{21}$$

where $Q_{ij}^{[*]} = \frac{-\epsilon_j}{\left(x_i^{[r]} - \zeta_j\right)\left(x_i^{[r]} - u_j^{[r]}\right)}$ and, from (20), $u_j^{[r]} - \zeta_j = O\left(\left|\left[\epsilon_j\right]^2\right|\right)$. Thus,

$$\epsilon_y = \frac{\epsilon_i \sum_{\substack{j=1 \\ j \neq i}}^{n} \epsilon_j^2 Q_{ij}^{[*]}}{1 + \epsilon_i \sum_{\substack{j=1 \\ j \neq i}}^{n} \epsilon_j^2 Q_{ij}^{[*]}}, \tag{22}$$

Assuming $|\epsilon_i| = |\epsilon_j| = \epsilon,$ we get

$$\epsilon_y = O\left(\left|[\epsilon]^3\right|\right). \tag{23}$$

In the second step, we have:

$$v_i^{[r]} - \zeta_i = x_i^{[r]} - \zeta_i - \frac{\left(\frac{2 \prod_{\substack{j=1 \\ j \neq i}}^{n} \left(x_i^{[r]} - x_j^{[r]}\right)}{\prod_{\substack{j=1 \\ j \neq i}}^{n} \left(x_i^{[r]} - x_j^{[r]}\right) - \prod_{\substack{j=1 \\ j \neq i}}^{n} \left(y_i^{[r]} - y_j^{[r]}\right)} \right)}{\left[\frac{h'\left(x_i^{[r]}\right)}{h\left(x_i^{[r]}\right)} - \sum_{\substack{j=1 \\ j \neq i}}^{n} \left(\frac{1}{\left(x_i^{[r]} - u_j^{[r]}\right)} \right) \right]}, \tag{24}$$

leading to:

$$\epsilon_v = \epsilon_i - \frac{\left(\frac{2 \prod_{\substack{j=1 \\ j \neq i}}^{n} \left(x_i^{[r]} - x_j^{[r]}\right)}{\prod_{\substack{j=1 \\ j \neq i}}^{n} \left(x_i^{[r]} - x_j^{[r]}\right) - \prod_{\substack{j=1 \\ j \neq i}}^{n} \left(y_i^{[r]} - y_j^{[r]}\right)} \right)}{\frac{1}{x_i^{[r]} - \zeta_i} + \sum_{\substack{j=1 \\ j \neq i}}^{n} \left(\frac{1}{\left(x_i^{[r]} - \zeta_j\right)} \right) - \sum_{\substack{j=1 \\ j \neq i}}^{n} \left(\frac{1}{\left(x_i^{[r]} - u_j^{[r]}\right)} \right)}, \tag{25}$$

$$\epsilon_v = \epsilon_i - \frac{\left(\dfrac{2\prod_{\substack{j=1\\j\neq i}}^{n}\left(x_i^{[r]}-x_j^{[r]}\right)}{\prod_{\substack{j=1\\j\neq i}}^{n}\left(x_i^{[r]}-x_j^{[r]}\right)-\prod_{\substack{j=1\\j\neq i}}^{n}\left(y_i^{[r]}-y_j^{[r]}\right)}\right)\epsilon_i}{1+\epsilon_i\sum_{\substack{j=1\\j\neq i}}^{n}\left(\dfrac{-u_j^{[r]}+\zeta_j}{\left(x_i^{[r]}-\zeta_j\right)\left(x_i^{[r]}-u_j^{[r]}\right)}\right)}, \tag{26}$$

$$\epsilon_v = \epsilon_i - \frac{\epsilon_i+\epsilon_i\sum\limits_{\substack{j=1\\j\neq i}}^{n}\left(\dfrac{-u_j^{[r]}+\zeta_j}{\left(x_i^{[r]}-\zeta_j\right)\left(x_i^{[r]}-u_j^{[r]}\right)}\right)-\left(\dfrac{2\epsilon_i}{1-\dfrac{\prod_{\substack{j=1\\j\neq i}}^{n}\left(y_i^{[r]}-y_j^{[r]}\right)}{\prod_{\substack{j=1\\j\neq i}}^{n}\left(x_i^{[r]}-x_j^{[r]}\right)}}\right)\epsilon_i}{1+\epsilon_i\sum\limits_{\substack{j=1\\j\neq i}}^{n}\left(\dfrac{-u_j^{[r]}+\zeta_j}{\left(x_i^{[r]}-\zeta_j\right)\left(x_i^{[r]}-u_j^{[r]}\right)}\right)}.$$

Assuming $|\epsilon_i| = |\epsilon_j| = \epsilon$, we have:

$$\epsilon_v = \frac{\epsilon_i+\epsilon_i\sum\limits_{\substack{j=1\\j\neq i}}^{n}\epsilon_j^2 Q_{ij}^{[*]}-\left(\dfrac{2\epsilon_i}{1-\dfrac{\prod_{\substack{j=1\\j\neq i}}^{n}\left(y_i^{[r]}-y_j^{[r]}\right)}{\prod_{\substack{j=1\\j\neq i}}^{n}\left(x_i^{[r]}-x_j^{[r]}\right)}}\right)\epsilon_i}{1+\epsilon_i\sum\limits_{\substack{j=1\\j\neq i}}^{n}\epsilon_j^2 Q_{ij}^{[*]}} = O\left(\left|\epsilon^3\right|\right). \tag{27}$$

This completes the proof.

3. Numerical results

The numerical results of the hybrid multiplicative calculus-based parallel scheme have clearly shown the efficiencies and effectiveness of the proposed method in solving problem (1) as well as its practical applicability. Results from the parallel schemes offer very valuable insights on error analysis, computational efficiency, as well as convergence rates, which are indispensable for evaluating the overall performance and impact of this study. In this section, we check the efficacy and robustness of the method by applying it to some engineering applications, using the termination criterion above implemented in Maple 18:

$$(i) \qquad e_i^{[r]} = \left\| x_i^{[r+1]} - x_i^{[r]} \right\| < \in = 10^{-18}, \tag{28}$$

where $e_i^{[r]}$ represents the residual error in the in the L_2-norm. We compare our method with the methods proposed by Petkovic et al. [25] $\left(\mathrm{ZM}_1^{[*]}\right)$, Rafiq et al. [26] $\left(\mathrm{ZM}_2^{[*]}\right)$, and Nedzhibov [27] $\left(\mathrm{ZM}_3^{[*]}\right)$, all of which have a convergence order of three.

Multiplicative calculus-based numerical techniques have potential uses in a variety of domains where complex behavior is frequently governed by nonlinear equations, including fluid dynamics, control theory, economics, and biological modeling. Fluid implementations in fluid dynamics can more accurately depict shock fronts and turbulence because they are more intuitive in terms of energy dissipation and scaling, which improves the stability and efficiency of computational fluid dynamics models. They can contribute to control theory by enhancing the benefits of adaptive control systems and nonlinear stability analysis. Rapid convergence toward stability and good accuracy are necessary for some real-time changes for applications, like autonomous cars and robotics.

Further, these techniques typically work well when traditional numerical schemes are failed to solve and also applied these multiplicative calculus-based schemes to biological and economic models that exhibit oscillatory or exponential growth, such as population dynamics and economic growth models. Here, we examine some engineering applications that demonstrate the consistency and reliability of newly developed method in comparison with classical methods.

3.1 Dynamical analysis of nonlinear frame structures: An eigenvalue problem

In engineering, the analysis of nonlinear frame structures [28] is essential for simulating the complex behavior of mechanical systems, buildings, bridges, and other structures subjected to dynamic loads. Material nonlinearities and significant deformations in these structures often go beyond the limitations of linear models. To ensure stability, safety, and performance in real-world scenarios, such as during earthquakes or under wind loads, it is crucial to understand their nonlinear behavior. Accurate analysis enables engineers to design structures that can withstand extreme stresses and prevent issues such as buckling or resonance, making nonlinear analysis vital for structural optimization and safety.

The governing equations for nonlinear frame structures are sixth-order nonlinear partial differential equations (PDEs) in both time and space, as given by:

$$E_1^{[*]} I_1^{[*]} \left(\frac{\partial^6 U(x,t)}{\partial x^6} + \frac{\partial^4 U(x,t)}{\partial x^4} \right) + \rho^{[*]} A^{[*]} \frac{\partial^2 U(x,t)}{\partial t^2} + \alpha^{[*]} U(x,t) + h_n(U(x,t)) = 0, \tag{29}$$

where $h_{[n]}(U(x,t))$ is a nonlinear functions.

After linearizing the given PDE, it becomes:

$$\begin{cases} E_1^{[*]} I_1^{[*]} \left(\dfrac{\partial^6 U(x,t)}{\partial x^6} + \dfrac{\partial^4 U(x,t)}{\partial x^4} \right) + \rho^{[*]} A^{[*]} \dfrac{\partial^2 U(x,t)}{\partial t^2} + \alpha^{[*]} U(x,t) \\ \qquad + h_{[n]}(U(x,t)) = 0. \end{cases} \tag{30}$$

Next, we transform Eq. (30) from the time domain to the frequency domain using:

$$\hat{U}(x,t) = \int_{-\infty}^{\infty} U(x,t)e^{-iUt}dt, \quad (31)$$

where U is the angular frequency. Applying Eqs. (31) to (30), we obtain:

$$\left\{ E_1^{[*]} I_1^{[*]} \left(\frac{d^6 \hat{U}(x,U)}{dx^6} + \frac{d^4 \hat{U}(x,U)}{dx^4} \right) + \rho^{[*]} A^{[*]} \hat{U}^2(x,U) + \alpha^{[*]} \hat{U}(x,U) = 0 \right. \quad (32)$$

Assuming $\hat{U}(x,U) = e^{\lambda x}$, where λ is a constant and $k \in \mathbb{C}$, we have:

$$E_1^{[*]} I_1^{[*]} \left(\lambda^6 e^{\lambda x} + \lambda^4 e^{\lambda x} \right) + \rho^{[*]} A^{[*]} U^2 e^{\lambda x} + \alpha^{[*]} e^{\lambda x} = 0. \quad (33)$$

By taking $e^{\lambda x} \neq 0$, we simplify the equation to:

$$h(\lambda) = E_1^{[*]} I_1^{[*]} \left(\lambda^6 + \lambda^4 \right) + \rho^{[*]} A^{[*]} U^2 + \alpha^{[*]}. \quad (34)$$

Using the following specific values:

$E_1^{[*]}$	200×10^9 Pa	(Young's modulus for steel)
$I_1^{[*]}$	$2 \times 10^{-6} \mathrm{m}^4$	(Moment of inertia of the cross-section)
$\rho^{[*]}$	$7850\,\mathrm{kg/m}^3$	(Density of steel)
$A^{[*]}$	$0.005\,\mathrm{m}^2$	(Cross-sectional area)
$\alpha^{[*]}$	$10^6/\mathrm{m}^4$	Stiffness constant
U	100 rad/s	Angular frequency

in Eq. (34), gives

$$h(\lambda) = \left(200 \times 10^9\right)\left(2 \times 10^{-6}\right)\left(\lambda^6 + \lambda^4\right) - 7850 \times 0.005U^2 + 10^6 = 0, \quad (35)$$

$$h(\lambda) = 4000\lambda^6 + 4000\lambda^4 - 607500 = 0, \quad (36)$$

or:

$$h(\lambda) = \lambda^6 + \lambda^4 - 151.875. \quad (37)$$

The eigenvalues of Eq. (37), correct to four decimal places, are: $\lambda_1 = -2.2409$, $\lambda_2 = -1.1154 - 2.062i, \lambda_3 = -1.1154 + 2.062i, \lambda_4 = 1.1154 - 2.062i, \lambda_5 = 1.1154 + 2.062i, \lambda_6 = -2.0627i$ For greater accuracy, we compute the eigenvalues using the parallel schemes, starting with initial values close to one decimal place. The results are presented in **Table 1**.

To assess global convergence, we used the following randomly generated initial guesses for the eigenvalues, as presented in **Table 2** and **Figures 1–3**:

Method	$e_1^{[6]}$	$e_2^{[6]}$	$e_3^{[6]}$	$e_4^{[6]}$	$e_5^{[6]}$	$e_6^{[6]}$
$ZM_1^{[*]}$	4.19×10^{-25}	5.15×10^{-14}	8.10×10^{-33}	1.76×10^{-44}	1.33×10^{-21}	4.19×10^{-35}
$ZM_2^{[*]}$	6.00×10^{-32}	1.19×10^{-32}	4.19×10^{-35}	1.08×10^{-27}	5.68×10^{-29}	6.00×10^{-42}
$ZM_3^{[*]}$	3.02×10^{-34}	9.17×10^{-31}	6.00×10^{-22}	3.02×10^{-24}	5.57×10^{-25}	3.02×10^{-34}
$ZM^{[*]}$	0.15×10^{-34}	2.01×10^{-45}	2.01×10^{-51}	0.15×10^{-44}	0.65×10^{-68}	0.15×10^{-54}

Table 1.
Numerical results of parallel schemes $ZM_1^{[*]} - ZM_3^{[*]}$ *and* $ZM^{[*]}$ *for solving frame structure engineering problems.*

$\lambda_i^{[*]}$	$[\lambda_1^{[0]}$	$\lambda_2^{[0]}$	$\lambda_3^{[0]}$	$\lambda_4^{[0]}$	$\lambda_5^{[0]}$	$\lambda_6^{[0]}]$
$\lambda_1^{[*]}$	[0.43	0.15	0.10	0.76	0.33	0.18]
$\lambda_2^{[*]}$	[0.03	0.19	0.19	0.08	0.68	0.17]
$\lambda_3^{[*]}$	[0.75	0.17	0.01	0.02	0.57	0.35]

Table 2.
Random initial eigenvalues for (37).

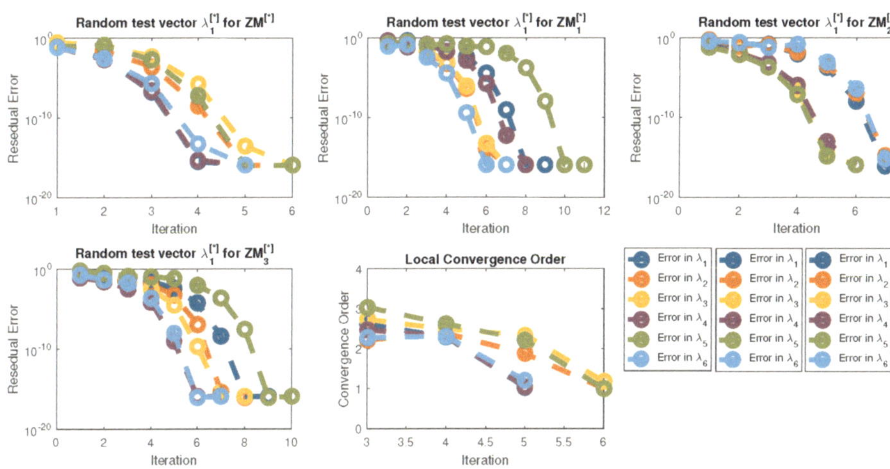

Figure 1.
Error graph of the parallel methods $ZM^{[*]}, ZM_1^{[*]}$*-*$ZM_3^{[*]}$ *for random initial Gaussian vectors* $\lambda_1^{[*]}$.

3.2 Stability and sensitivity analysis

To evaluate the convergence of our approach compared to existing methods, we analyzed the results of the numerical schemes based on random initial eigenvalues, as shown in **Table 4**. This table includes the number of iterations (n), the number of convergence operations (Op-con), the percentage of convergence (Per-con), the maximum errors $\left(\text{Max-E}_{\lambda_1^{[*]}} - \text{Max-E}_{\lambda_3^{[*]}}\right)$ for random initial values, and the computing time in seconds (CPU-time).

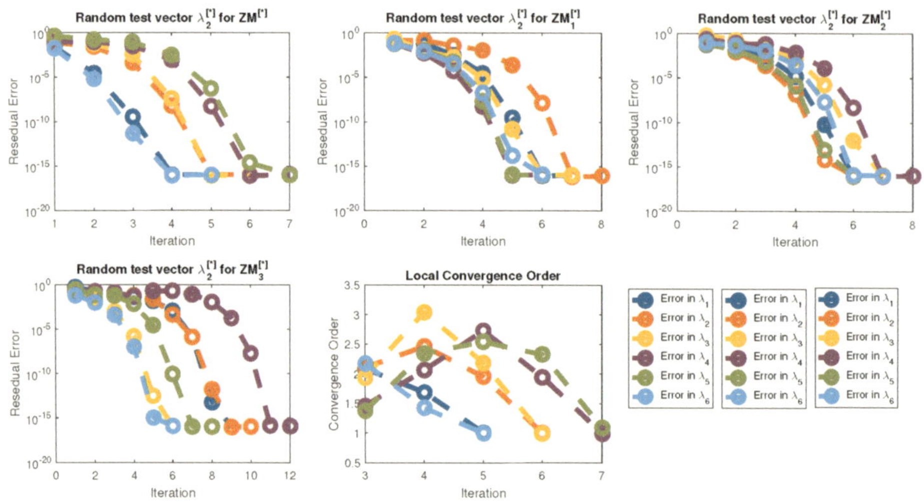

Figure 2.
Error graph of the parallel methods $ZM^{[*]}, ZM_1^{[*]}$-$ZM_3^{[*]}$ *for random initial Gaussian vectors* $\lambda_2^{[*]}$.

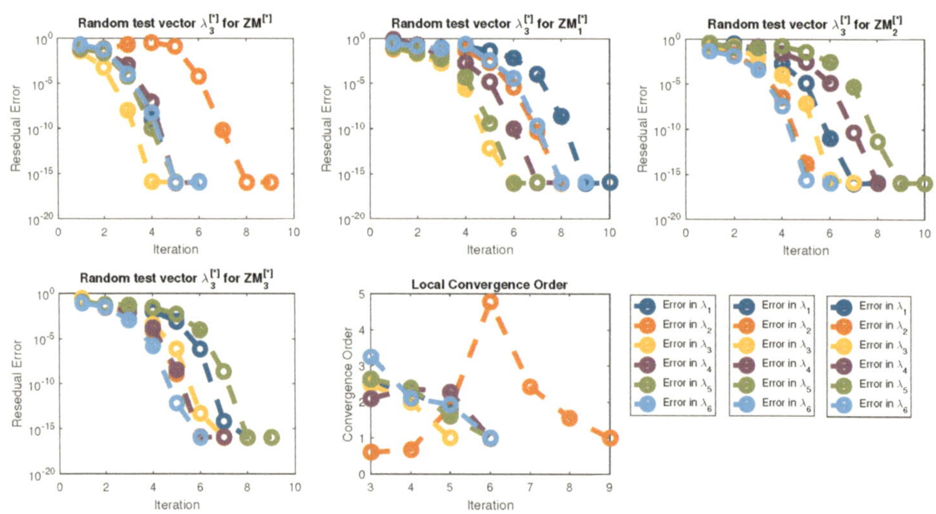

Figure 3.
Error graph of the parallel methods $ZM^{[*]}, ZM_1^{[*]}$-$ZM_3^{[*]}$ *for random initial Gaussian vectors* $\lambda_3^{[*]}$.

In terms of iterations, computing time, percentage convergence, and maximum error for random initial guessed eigenvalues, **Tables 3** and **4** clearly show that our methods outperform those currently used in the literature. **Table 4** demonstrates that our newly developed procedure is more consistent and reliable than other methods. We then examine how the frame structure problem (1) is solved using the calculated eigenvalues. The final solution is obtained by applying the inverse of (31) to the frequency-domain solution, resulting in:

Error	$e_1^{[0]}$	$e_2^{[0]}$	$e_3^{[0]}$	$e_4^{[0]}$	$e_5^{[0]}$	$e_6^{[0]}$
Numerical results for $\lambda_1^{[*]}$ used to determine all eigenvalues of Eq. (37).						
$ZM_1^{[*]}$	1.11×10^{-3}	5.15×10^{-5}	8.10×10^{-3}	1.76×10^{-4}	1.33×10^{-11}	1.11×10^{-3}
$ZM_2^{[*]}$	0.08×10^{-5}	1.19×10^{-4}	4.19×10^{-5}	1.08×10^{-7}	5.68×10^{-19}	0.08×10^{-5}
$ZM_3^{[*]}$	0.05×10^{-1}	9.17×10^{-6}	6.00×10^{-6}	3.02×10^{-8}	5.57×10^{-15}	0.05×10^{-11}
$ZM^{[*]}$	2.01×10^{-5}	2.77×10^{-17}	2.92×10^{-22}	1.05×10^{-3}	0.68×10^{-25}	2.01×10^{-10}
Numerical results for $\lambda_2^{[*]}$ used to determine all eigenvalues of Eq. (37).						
$ZM_1^{[*]}$	1.11×10^{-3}	0.45×10^{-1}	0.07×10^{-2}	1.91×10^{-13}	0.03×10^{-11}	1.51×10^{-3}
$ZM_2^{[*]}$	0.08×10^{-5}	0.04×10^{-12}	4.09×10^{-14}	0.08×10^{-5}	0.65×10^{-15}	0.08×10^{-5}
$ZM_3^{[*]}$	1.91×10^{-3}	1.71×10^{-13}	0.05×10^{-11}	1.03×10^{-2}	0.50×10^{-15}	0.05×10^{-5}
$ZM^{[*]}$	0.08×10^{-25}	1.08×10^{-25}	2.01×10^{-31}	0.15×10^{-24}	0.65×10^{-18}	2.01×10^{-11}

Table 3.
*The numerical results using these random initial guesses are presented in **Tables 3** and **4**.*

Method	n	CPU-time	Per-con (%)	Op-con	Max-$E_{\lambda_1^{[*]}}$	Max-$E_{\lambda_2^{[*]}}$	Max-$E_{\lambda_3^{[*]}}$
$ZM_1^{[*]}$	17	0.4315	37	234	0.05×10^{-24}	0.15×10^{-21}	0.15×10^{-27}
$ZM_2^{[*]}$	15	0.4359	32	434	0.45×10^{-11}	1.15×10^{-12}	7.05×10^{-13}
$ZM_3^{[*]}$	11	0.3243	52	285	2.15×10^{-15}	3.05×10^{-13}	0.07×10^{-10}
$ZM^{[*]}$	09	0.2137	69	187	0.32×10^{-16}	0.04×10^{-11}	0.09×10^{-15}

Table 4.
Random initial eigenvalues for finding all eigenfunctions simultaneously.

$$U(x,t) = \sum_{i=1}^{6} T_i^{[*]}(t)\,\vartheta_i^{[*]}(x) \tag{38}$$

where $T_i^{[*]}(t) = A_i^{[*]} \cos\left(U_i t + \vartheta_i^{[*]}\right)$ represents the time-dependent part of the ù solution for the i-th mode, with $A_i^{[*]}$ being the amplitude and U_i the frequency associated with the i-th eigenvalue λ_i. The spatial mode shape, $\vartheta_i^{[*]}(x)$, depends on the eigenvalue λ_i. This mode is typically sinusoidal, for example:

$$\vartheta_i^{[*]}(x) = \sin\left(\frac{i\pi x}{L^{[*]}}\right), \tag{39}$$

where $L^{[*]}$ is the length of the frame structure, which is 10 m for the natural frequency $U_i = \sqrt{\frac{\lambda_i E_1^{[*]} I_1^{[*]}}{\rho^{[*]} A^{[*]}}}$. The value of λ_i determines the various modes of vibration in the frame structure. The forward-traveling waves correspond to $\lambda_2, \lambda_4, \lambda_6,$ while the backward-traveling waves correspond to $\lambda_1, \lambda_3, \lambda_5$. Complex oscillations arise from the

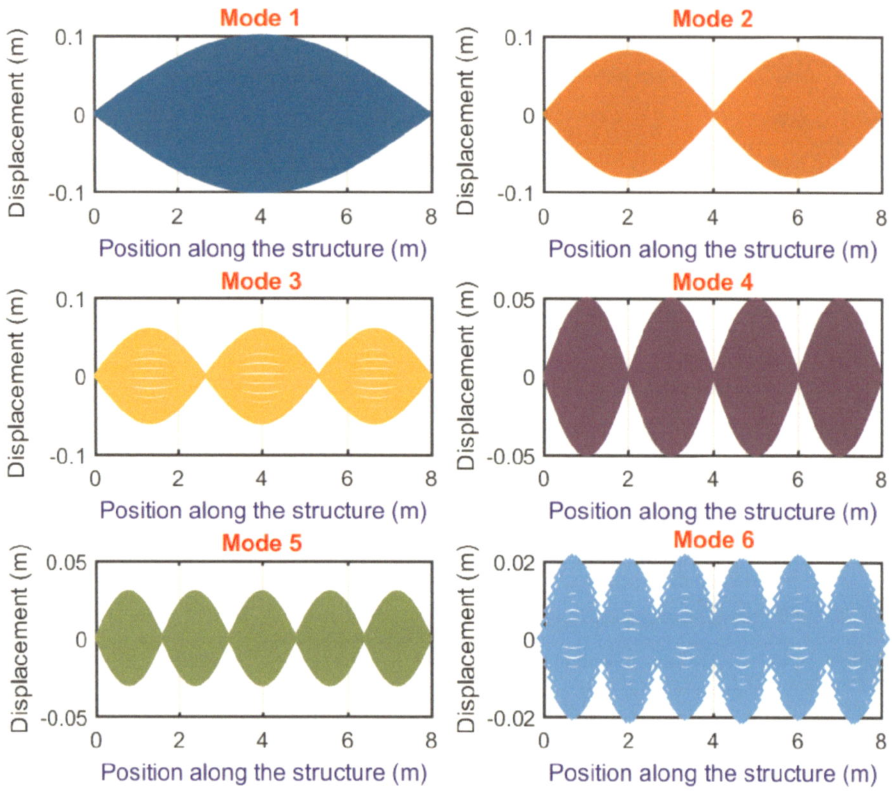

Figure 4.
Dynamical analysis of the frame structure modes for low and higher values of λ.

negative values of λ_i, indicating oscillations in the opposite direction. The magnitude of the eigenvalue indicates the frequency of the oscillations, with larger values corresponding to higher-frequency modes. Low-frequency modes (λ_5, λ_6) are associated with longer wavelengths and smoother, more widespread frame deformations, while high-frequency modes (λ_1, λ_2) produce localized, faster oscillations with shorter wavelengths.

3.3 Nature of the solutions

- Low-frequency modes corresponding to λ_5, λ_6 : These modes are associated with the global deformation of the entire structure. The displacement profile shows smooth transitions and bending of the frame along its full length. These modes are typical in real-world situations where overall bending is the primary concern, such as in bridges or large structural frames.

- High-frequency modes corresponding to λ_1, λ_2 : These modes involve localized, higher-order deformations in specific areas of the frame. As the frequency increases, the displacement profile exhibits wave-like oscillations with many points of inflection along the frame length. In practice, these modes may correspond to vibrations caused by impacts or high-energy disturbances.

DOI: http://dx.doi.org/10.5772/intechopen.1008844

- Superposition mode: The final solution is a combination of all mode shapes and their time-dependent oscillations. In reality, the observed displacement is a result of all these modes interacting simultaneously.

4. Conclusion

In this chapter, we introduced a new multiplicative calculus-based parallel technique for solving nonlinear eigenvalue problems. We conducted an in-depth analysis of the dynamical behavior of nonlinear frame structures, using the eigenvalue problem to assess the stability and consistency of the newly developed scheme. This scheme was compared to existing parallel methods with the same convergence order (three). As demonstrated in **Tables 1–3** and **Figures 1–4**, our proposed method, $ZM^{[*]}$, outperforms existing methods $ZM_1^{[*]} - ZM_3^{[*]}$ in terms of iteration count, residual error, and computational convergence rates, even with random initial starting values.

Future extensions will include a number of enhancements, such as high-order multiplicative approaches, the examination of applications in higher-dimensional systems, and combinations with machine learning to give hybrid models [29], improved convergence, and accuracy. Furthermore, research into the fractal properties of those schemes would provide insights into managing chaotic and sensitive systems such as financial markets, fluid mechanics, epidemic models, biomedical engineering applications, and a strong software framework to as many universals as possible, allowing the methods to be applied to a wide range of scientific and engineering applications.

Conflict of interest

The authors declare no conflict of interest.

Author details

Mudassir Shams[1,2*†] and Bruno Carpentieri[1†]

1 Faculty of Engineering, Free University of Bozen-Bolzano, Bolzano, Italy

2 Department of Mathematics, Faculty of Arts and Science, Balıkesir University, Balıkesir, Turkey

*Address all correspondence to: bruno.carpentieri@unibz.it

† These authors contributed equally.

References

[1] Dautray R, Lions JL. Mathematical Analysis and Numerical Methods for Science and Technology: Volume 1 Physical Origins and Classical Methods. Berlin, Germany: Springer Science and Business Media; 2012

[2] Hamming R. Numerical Methods for Scientists and Engineers. North Chelmsford, Massachusetts, USA: Courier Corporation; 2012

[3] McLachlan AD. A variational solution of the time-dependent Schrodinger equation. Molecular Physics. 1964;**8**(1): 39-44

[4] Bilgehan B, Eminağa B, Riza M. New solution method for electrical systems represented by ordinary differential equation. Journal of Circuits, Systems and Computers. 2016;**25**(02): 1650011

[5] Anderson DR, Ulness DJ. Newly defined conformable derivatives. Advances in Dynamical Systems and Applications. 2015;**10**(2):109-137

[6] Kraka E, Quintano M, La Force HW, Antonio JJ, Freindorf M. The local vibrational mode theory and its place in the vibrational spectroscopy arena. The Journal of Physical Chemistry A. 2022; **126**(47):8781-8798

[7] Losanno D, Londono JM, Zinno S, Serino G. Effective damping and frequencies of viscous damper braced structures considering the supports flexibility. Computers and Structures. 2018;**207**:121-131

[8] Hommes C. Behavioral Rationality and Heterogeneous Expectations in Complex Economic Systems. Cambridge, United Kingdom: Cambridge University Press; 2013

[9] Funke J. Solving complex problems: Exploration and control of complex systems. In: Complex Problem Solving. New York, NY, USA: Psychology Press; 2014. pp. 185-222

[10] Coates V, Farooque M, Klavans R, Lapid K, Linstone HA, Pistorius C, et al. On the future of technological forecasting. Technological Forecasting and Social Change. 2001;**67**(1):1-17

[11] Galambos TV, Surovek AE. Structural Stability of Steel: Concepts and Applications for Structural Engineers. John Wiley & Sons, Inc. 2008

[12] Dehuri S, Cho SB. A comprehensive survey on functional link neural networks and an adaptive PSO–BP learning for CFLNN. Neural Computing and Applications. 2010;**19**:187-205

[13] Ong H. Linear transformations on matrices: The invariance of generalized permutation matrices—III. Linear Algebra and its Applications. 1976;**15**(2): 119-151

[14] Sorensen HK. Niels Henrik Abel and the Theory of Equations. Aarhus: Appendix of Progress Report, Institut for Videnskabshistorie, Aarhus Universitet; 1999

[15] Shams M, Carpentieri B. An efficient and stable caputo-type inverse fractional parallel scheme for solving nonlinear equations. Axioms. 2024;**13**(10):671

[16] Shams M, Carpentieri B. Q-analogues of parallel numerical scheme based on neural networks and their engineering applications. Applied Sciences. 2024;**14**(4):1540

[17] Shams M, Carpentieri B. Efficient inverse fractional neural network-based

simultaneous schemes for nonlinear engineering applications. Fractal and Fractional. 2023;**7**(12):849

[18] Bashirov AE, Kurpınar EM, Özyapıcı A. Multiplicative calculus and its applications. Journal of Mathematical Analysis and Applications. 2008;**337**(1): 36-48

[19] Yener G, Emiroglu I. A q-analogue of the multiplicative calculus: Q-multiplicative calculus. Discrete and Continuous Dynamical Systems – Series S. 2015;**8**:1435-1450

[20] Özyapıcı A, Sensoy ZB, Karanfiller T. Effective root-finding methods for nonlinear equations based on multiplicative calculi. Journal of Mathematics. 2016;**2016**(1):8174610

[21] Riza M, Aktöre H. The Runge–Kutta method in geometric multiplicative calculus. LMS Journal of Computation and Mathematics. 2015;**18**(1):539-554

[22] Singh G, Bhalla S, Behl R. Higher-order multiplicative derivative iterative scheme to solve the nonlinear problems. Mathematical and Computational Applications. 2023;**28**(1):23

[23] Shams M, Rafiq N, Carpentieri B, Ahmad Mir N. A new approach to multiroot vectorial problems: Highly efficient parallel computing schemes. Fractal and Fractional. 2024;**8**(3):162

[24] Nishani HPS, Weerakoon S, Fernando TGI, Liyanage M. Weerakoon-Fernando method with accelerated third-order convergence for systems of nonlinear equations. International Journal of Mathematical Modelling and Numerical Optimisation. 2018;**8**(3): 287-304

[25] Petković MS, Rančić LZ. On the guaranteed convergence of a cubically convergent Weierstrass-like root-finding method. International Journal of Computer Mathematics. 2015;**92**(6): 1303-1312

[26] Rafiq N, Mir NA, Yasmin N. Some two-step simultaneous methods for determining all the roots of a non-linear equation. Life Sciences. 2013;**10**:54-59

[27] Nedzhibov GH, Petkov MG. On a family of iterative methods for simultaneous extraction of all roots of algebraic polynomial. Applied Mathematics and Computation. 2005; **162**(1):427-433

[28] Bendat JS, Piersol AG. Engineering Applications of Correlation and Spectral Analysis. New York; John Wiley & Sons; 1980

[29] Shams M, Rafiq N, Kausar N, Agarwal P, Park C, Mir NA. On iterative techniques for estimating all roots of nonlinear equation and its system with application in differential equation. Advances in Difference Equations. 2021; **2021**(1):1-18

Chapter 6

Eigen-Analysis of Multi-Agent Systems and Large Scale Systems Using Data Driven and Machine Learning Algorithms

Kenneth McDonald, Zhihua Qu and Azwirman Gusrialdi

Abstract

Eigenvalue analysis is central in stability analysis and control design of linear dynamic systems. While eigen-analysis is a standard tool, determining eigenvalues of multi-agent systems and/or interconnected dynamical systems remains challenging due to the sheer size of such systems, changes of their topology, and limited information about subsystems' dynamics. In this chapter, a set of scalable, data-driven estimation and machine learning algorithms are presented to determine eigenvalue(s) and in turn stability of such large-scale complex systems. We begin with distributed algorithms that estimate all the eigenvalues of multi-agent cooperative systems, where their subsystems are modeled as a single integrator and interconnected by local communication networks. The algorithms are then extended to the data-driven version that estimate the dominant eigenvalues of large-scale interconnected systems with unknown dynamical model. Subsequently, we study input-output stability of subsystems and extend eigen-analysis to investigation of passivity shortage using the input-output data. This analysis is then further extended to machine learning algorithms by which stability properties of unknown subsystems can be learned. These results are illustrated by examples.

Keywords: stability, eigenvalues, multi-agent systems, consensus algorithm, input-output stability, passivity and passivity shortage, machine learning

1. Introduction

Eigen-analysis is a fundamental and well known concept in linear algebra and linear systems theory. Technically, it deals with eigenvalues and eigenvectors of matrices or linear transformations. When applied to linear systems, eigenanalysis quantifies characteristics of dynamic responses, reveals such properties as stability and robustness, and provides closed-form solutions. As applications move toward multi-agent systems, large scale networked systems, and machine learning, eigen-analysis remains to be an effective approach for qualitative and quantitative analyses. This chapter aims to illustrate this fact by focusing upon a few of contemporary topics.

This chapter begins with a summary of most foundational results in Section 2. The section contains the following specific results. Section 2.1 provides eigenanalysis of linear time-invariant systems, their solution, and their stability by summarizing the basic results from [1, 2]. Section 2.2 introduces Lyapunov direct method [3] and presents quadratic Lyapunov analysis in terms of eigenvalues. The Lyapunov direct method applies to both linear and nonlinear dynamic systems. One way to bridge analyses of linear and nonlinear dynamic systems is to parameterize their stability analysis. To this end, Section 2.3 introduces dissipativity theory [4], which allows for the classification of linear and nonlinear systems into passive and passivity-short systems, providing a unified framework for their analyses. In Section 2.4, eigen-analysis is applied to investigate stability analysis of multi-agent systems, including cooperative and networked systems [5], and the result enables a scalable and modular design [6] of networked control systems in terms of two key parameters on system's input-output relationship. In Section 2.5, input-output relationship is explicitly derived for linear time invariant systems for the purpose of performing machine learning as well as the subsequent stability analysis and control design.

Section 3 explores the eigen-analysis of multi-agent and large scale systems, with a particular focus on systems that may have unknown models. The first part addresses the problem of distributed eigenvalue estimation in multiagent systems, where each agent has access only to local information. While various distributed algorithms have been proposed to tackle this issue (e.g., [7–10]), existing approaches such as the power iteration [7, 8] and consensus-based algorithm [9] are typically limited to estimating only the dominant eigenvalues. Even algorithms that can estimate all eigenvalues, such as [10], are often restricted to specific types of matrices, such as rowstochastic ones. To address these limitations, Section 3.1 presents distributed algorithms capable of estimating all eigen-values for any irreducible matrix, broadening the applicability of eigenvalue estimation methods. The discussion then extends to the challenge of distributed dominant eigenvalue estimation for unknown linear time-invariant systems within autonomous, large scale systems. Existing data-driven techniques, including dynamic mode decomposition [11, 12], power iteration [13], prony method [14], distributed optimization-based approach [15], and Hankel matrix [16], each face drawbacks. These include centraliza-tion, applicability only to Laplacian matrices, or limitations to matrices with distinct eigenvalues. To overcome these challenges, two distributed datadriven algorithms are presented in Section 3.2, which estimate eigenvalues by learning local models and apply-ing model reduction techniques, making them suitable for handling large scale models.

Different from Section 3, Section 4 addresses stability analysis and control design through machine learning. Specifically, a data driven algorithm is presented to learn the two key parameters. To demonstrate connection and effectiveness, the two parameters are calculated using both approaches of eigenanalysis and machine learning. The machine learning approach bypasses the step of model identification and hence is more direct and efficient in design and analysis. Combined with the results in Section 2.4, machine learning can be applied to multi-agent systems as well as large scale systems.

2. Preliminaries

2.1 Analysis of linear time-invariant systems and their stability

Eigenvalue analysis is both fundamental and straightforward to investigate stabil-ity of linear time-invariant systems of form

$$\dot{x} = Ax + Bu, \quad y = Cx + Du, \tag{1}$$

where $x \in \Re^n$ is the state, $u \in \Re^m$ is the control input, and $y \in \Re^l$ is the output. Defining the *matrix exponential function* e^{At} as

$$e^{At} = \sum_{j=0}^{\infty} \frac{1}{j!} A^j t^j \tag{2}$$

we know that e^{At} has the property of

$$\frac{d}{dt} e^{At} = Ae^{At} = e^{At} A.$$

Hence, the solution to system (1) is

$$\dot{x}(t) = e^{At} x(0) + \int_0^t e^{A(t-\tau)} Bu(\tau) d\tau \tag{3}$$

in which $e^{A(t-t_0)}$ is also called the state transition matrix.

Given matrix A, its eigenvalues λ_i and eigenvectors v_i are defined as

$$Av_i = \lambda_i v_i, \quad i = 1, \cdots, n.$$

Matrix $A \in \Re^{n \times n}$ has n eigenvalues and, if the number of eigenvectors is r but less than n, it always has n eigenvectors and generalized eigenvectors. Assembling these eigenvectors and generalized eigenvectors into matrix S, we have

$$S^{-1}AS = J, \quad J = \begin{bmatrix} J_1 & & \\ & \ddots & \\ & & J_r \end{bmatrix}, \quad J_i = \begin{bmatrix} \lambda_i & 1 & 0 & \cdots \\ 0 & \ddots & \ddots & \ddots \\ \vdots & \ddots & \ddots & 1 \\ 0 & \cdots & 0 & \lambda_i \end{bmatrix} \in \Re^{n_i \times n_i},$$

where J is the Jordan canonical form with diagonal blocks J_i, and n_i is the geometric multiplicity of eigenvalue λ_i.

It follows from the structure of J and the Taylor series expansion in (2) that

$$e^{At} = S \begin{bmatrix} e^{J_1 t} & & \\ & \ddots & \\ & & e^{J_l t} \end{bmatrix} S^{-1}, \quad e^{J_i t} = \begin{bmatrix} e^{\lambda_i t} & te^{\lambda_i t} & \cdots & \frac{t^{n_i - 1}}{(n_i - 1)!} e^{\lambda_i t} \\ 0 & \ddots & \ddots & \vdots \\ \vdots & \vdots & \ddots & te^{\lambda_i t} \\ 0 & 0 & \cdots & e^{\lambda_i t} \end{bmatrix}. \tag{4}$$

Therefore, the following necessary and sufficient conditions can be concluded from (3) and (4):

i. System (1) with $u = 0$ is Lyapunov stable if and only if none of its eigenvalues is in the right open half plane and those on the imaginary axis are of geometrical multiplicity one.

ii. System (1) with $u = 0$ is asymptotically stable if and only if matrix A is Hurwitz, i.e., its eigenvalues are all in the left open half plane.

iii. System (1) is input-to-state stable if and only if it is asymptotically stable.

iv. System (1) with $u = 0$ is exponentially stable if and only if it is asymptotically stable.

If A is not Hurwitz, control u can be designed to achieve stability when pair $\{A, B\}$ is controllable.

2.2 Stability analysis of linear time-varying systems

Consider linear time-varying system:

$$\dot{x} = A(t)x + B(t)u, \quad y = C(t)x + D(t)u \tag{5}$$

where $x \in \Re^n$ is the state, $u \in \Re^m$ is the input, and $y \in \Re^p$ is the output. It is known that pointwise eigenvalues of time-varying matrix $A(t)$ being in the left open half plan *do not* imply stability of system (5), and a counterexample can be found in [3]. Instead, eigen-analysis of linear time varying system should be performed through the Lyapunov direct method.

Consider the autonomous system:

$$\dot{x} = A(t)x, \quad x \in \Re^n. \tag{6}$$

For linear systems, Lyapunov function can always be chosen to be a quadratic function of form

$$V(x,t) = x^T P(t)x,$$

where $P(t)$ is a symmetric matrix. Its time derivative along trajectories of system (6) is also quadratic as

$$\dot{V}(x,t) = -x^T Q(t)x,$$

where $P(t)$ and $Q(t)$ are related by the so-called differential Lyapunov equation

$$\dot{P}(t) = -A^T(t)P(t) - P(t)A(t) - Q(t). \tag{7}$$

To determine whether system (6) is asymptotically stable or not, the following three-step backward process needs to be applied: (a) choose $Q(t)$ to be symmetric, uniformly bounded (in the sense that the maximum eigenvalue is uniformly bounded from above as $\lambda_{\max}(Q(t)) \le \bar{c} < \infty$), and positive definite (in the sense that in the sense that the minimum eigenvalue is uniformly above zero as $\lambda_{\min}(Q(t)) \ge \underline{c} > 0$). The simplest choice is $Q(t) = I$. (b) Solve $P(t)$ from Eq. (7). (c) System (6) is asymptotically stable if and only if solution $P(t)$ is uniformly bounded and positive definite. Should matrix $A(t)$ be constant, Eq. (7) is algebraic, and solution P is constant.

2.3 Analysis of nonlinear systems

Consider the following nonlinear affine system

$$\dot{x} = f(x,t) + g(x,t)u, \quad y = h(x,t), \tag{8}$$

where $x \in \Re^n, u \in \Re^m$, and $y \in \Re^l$ are the state, input, and output, respectively. System (8) includes system (6) as a special case. Stability of system (8) can be investigated using Lyapunov direct method, as in Section 2.2. Lyapunov function takes the general form of $V(x,t)$ and have the following time derivative along the trajectories of system (8):

$$\dot{V} = \frac{\partial V}{\partial t} + \left(\frac{\partial V}{\partial x}\right)[f(x,t) + g(x,t)u]. \tag{9}$$

To show asymptotic stability under $u = 0$, one need to choose a positive definite function $\eta(x)$ and solve for Lyapunov function $V(x,t)$ from the following partial differential equation:

$$\frac{\partial V}{\partial t} + \left(\frac{\partial V}{\partial x}\right)f(x,t) = -\eta(x). \tag{10}$$

Stability can be determined by checking whether solution $V(x,t)$ is both positive definite and decrescent (i.e., upper and lower bounded by positive definite functions $\gamma_1(x), \gamma_2(x)$ as $\gamma_1(x) \le V(x,t) \le \gamma_2(x)$).

To investigate input-output relationship of linear and nonlinear systems, we can use the dissipativity theory. System (8) is said to be dissipative with respect to a positive semi-definite (p.s.d.) storage function $V(x)$ and a supply rate function $\Phi(u,y)$ if $V(0) = 0$ and if, for all $x_0 \in \mathcal{X}$,

$$V(x(\infty)) - V(x(0)) \le \int_0^t \Phi(u(\tau), y(\tau))d\tau. \tag{11}$$

As a Lyapunov function, the so-called storage function represents a broader concept of the energy stored within the system. Consequently, the above inequality implies that the stored energy at any given time, as described by the storage function, is always less than or equal to the total energy supplied to the system, including the initial energy.

Should the supply function be quadratic, inequality (11) is satisfied with

$$\Phi(u,y) = -\eta(x) + u^T y - \frac{\epsilon}{2}\|u\|^2 - \frac{\rho}{2}\|y\|^2, \tag{12}$$

where $\eta(x)$ is a positive semi-definite function. If the storage function is positive definite, it is a Lyapunov function, and system (8) is asymptotically stable. Depending upon the values of parameters ϵ and ρ, several sub-classes of dissipativity are given in **Table 1**.

It is well known that passive linear systems must be of relative degree 0 or 1, Lyapunov stable, and also minimum phase (or inversely Lyapunov stable).

This means that most of the stable systems are not passive. Since the engineered systems, such as teleoperation of an n-link robot [17] and synchronous generator,

Sub-classes	ϵ	ρ
Passive	$\epsilon, \rho \geq 0$	
Input strictly passive (ISP)	>0	≥ 0
Output strictly passive (OSP)	≥ 0	>0
Input-output strictly passive (IOSP)	>0	>0
Passivity short (PS)	either $\epsilon < 0$ or $\rho < 0$	
Input-feedforward passivity short (IFPS)	<0	≥ 0
Output-feedback passivity short (OFPS)	≥ 0	<0

Table 1.
Sub-classes of dissipativity and their parameter values.

would have bounded outputs when their inputs are bounded, these systems are typically input-feedforward passivity short (IFPS). It is known that, with an appropriate self-feedback control, a stabilizable linear system can achieve the IFPS property [18]. It is shown in the rest of the chapter that both passivity and passivity shortage enable modular analysis/design of complex systems, indirect eigen-analysis using input-output data, and machine learning.

2.4 Multi-agent systems and cooperative networked systems

A multi-agent system consists of cooperative agents with simple dynamics:

$$\dot{y}_i = u_i, \quad i \in \mathcal{N}, \tag{13}$$

where $\mathcal{N} = \{1, \cdots, n\}$ is the set of agents. These agents communicate through a local communication network of graph $\mathcal{G} = (\mathcal{N}, \mathcal{E})$, where $\mathcal{E} \subseteq \mathcal{N} \times \mathcal{N}$ is the edge set and $e_{ij} \in \mathcal{E}$ implies that y_j is sent by the jth agent to the ith agent, or $j \in \mathcal{N}_i$ with $\mathcal{N}_i \subset \mathcal{N}$ is the ith agent's neighbor set. Graph $\mathcal{G}$ is generally directed, i.e., $e_{ij} \in \mathcal{E}$ does not necessarily mean $e_{ji} \in \mathcal{E}$ or vice versa. Graph $\mathcal{G}$ is said to be strongly connected if every node is connected by directed edges to any other node. These local interactions enable the following cooperative consensus protocol:

$$u_i = \sum_{j \in \mathcal{N}_i} w_{ij} \left(y_j - y_i \right), \tag{14}$$

where $w_{ij} > 0$ are weights. The resulting cooperative system can be expressed as

$$\dot{y} = -Ly, \tag{15}$$

where

$$L = \left[L_{ij} \right], \quad L_{ij} = \begin{cases} 0 & \text{if } j \neq i \text{ and } j \notin \mathcal{N}_i \\ -w_{ij} & \text{if } j \neq i \text{ and } j \in \mathcal{N}_i \\ \sum\limits_{j \in \mathcal{N}_i} w_{ij} & \text{if } j = i \end{cases} \tag{16}$$

is the so-called Laplacian. Laplacian L is always Lyapunov stable, $\lambda_1(L) = 0$ is the eigenvalue(s) closest to the imaginary axis, and the corresponding eigenvector is the vector of 1 s. Should $\mathcal{G}$ be strongly connected, $\lambda_1(L) = 0$ is unique, and all of y_i converge to the same consensus value. If $\mathcal{G}$ is undirected and connected, Laplacian L can be symmetric (with $w_{ij} = w_{ji}$), and its eigenvalues values $\lambda_i(L)$ are all real and the property that

$$0 = \lambda_1(L) < \lambda_2(L) \leq \cdots \leq \lambda_n(L). \tag{17}$$

While multi-agent systems are easy to analyze, networked systems from applications often have heterogeneous dynamics in the form of

$$\dot{z}_i = f_i(z_i, u_i), \quad y_i = h_i(z_i), \tag{18}$$

where $z_i \in \mathfrak{R}^{n_i}$ and $y_i \in \mathfrak{R}^l$ are the state and the output of the ith system, respectively. System (18) includes multi-agent system (13) as a special case. Instead of assembling all the dynamics from system (18) and analyzing the stability together, one can use the concepts of passivity and passivity shortage to perform modular analysis and design. To this end, first assume that system (18) has storage function $V_i(z_i)$ and the following dissipativity property:

$$\dot{V}_i \leq y_i^T u_i - \epsilon_i u_i^T u_i - \rho_i y_i^T y_i \tag{19}$$

where ϵ_i and ρ_i can assume values according to **Table 1**. Next, assume that Laplace L is symmetric and connected, and revise the consensus protocol (14) by incorporating cooperative control gain $\kappa > 0$ as

$$u_i = \kappa \sum_{j=1}^{n} w_{ij} \left(y_j - y_i \right). \tag{20}$$

Then, choosing the following Lyapunov function

$$V = \sum_{i=1}^{n} V_i, \tag{21}$$

and taking the time derivative along the trajectories of system (18) under control (20) yield

$$\begin{aligned} \dot{V} &\leq \sum_{i=1}^{n} \left[y_i^T u_i - \epsilon_i u_i^T u_i - \rho_i y_i^T y_i \right] \\ &= -y^T Q y, \end{aligned} \tag{22}$$

where

$$Q = \kappa L + \kappa^2 \operatorname{diag}\{\epsilon_i\} L^2 + \operatorname{diag}\{\rho_i\} I. \tag{23}$$

It is obvious that the overall system of heterogeneous subsystems reaches consensus if matrix Q has the property of $\lambda_{min}(Q) \geq 0$. This is ensured if the following scalar, quadratic inequality admits positive solution κ:

$$\kappa\lambda_2(L) + \kappa^2\left(\min\left\{0, \min_i \epsilon_i\right\}\right)\lambda_n^2(L) + \left(\min_i \rho_i\right) \geq 0. \tag{24}$$

A quick analysis of the above inequality reveals the following stability property for cooperative networked systems:

i. If all the systems in (18) are passive, consensus can be achieved for any $\kappa \in (0, \infty)$.

ii. If all the systems in (18) are either passive or IFPS, there exists $\overline{\kappa} > 0$ such that consensus can be achieved for any $\kappa \in (0, \overline{\kappa})$.

iii. If all the systems in (18) are either passive or OFPS, there exists $\underline{k} > 0$ such that consensus can be achieved for any $\kappa \in (\underline{\kappa}, \infty)$.

If the systems in (18) are a mixture of being passive, IFPS, and OFPS, consensus can still be achieved by changing common gain κ into individual gains κ_i. These results show that, for cooperative consensus control (20) or its individual gain version, choices of cooperative control gains are modular and plug-and-play in spite of heterogeneous dynamics (18).

2.5 Data-driven modeling

For linear systems, data-driven modeling refers to the approach of building models based solely on input-output data, without requiring explicit knowledge of the system's internal dynamics or parameters. This method is particularly useful when the system is a black or gray box, or its internal parameters are inaccessible. In this section, input-output relationship is established using discretization so that machine learning of key parameters can be done later in Section 4.

Consider linear system (1). Its discretized version is: given $x(0) = 0$ and sampling period T_s,

$$x_{l+1} = A_d x_l + B_d u_l, \quad y_l = Cx_l + Du_l, \quad A_d = e^{AT_s}, \quad B_d = A^{-1}(A_d - I)B. \tag{25}$$

To further represent the system in terms of inputs and outputs [19], consider the representation given by

$$y_l = \sum_{\tau=0}^{l} g_\tau u_{l-\tau}, \quad g_0 = D, \quad g_\tau = CA_d^{\tau-1}B_d, \forall \tau > 0. \tag{26}$$

Passive systems have relative degree r to be either 0 or 1. On the other hand, passivity short systems may have $r \geq 1$, causing the convoluted matrix G_d to becomes singular. This can be avoided by utilizing its reduced order representation as needed. Let $Y = G_d U$, where N is sufficiently large,

$$Y = \begin{bmatrix} y_r & y_{r+1} & \cdots & y_{N-1} \end{bmatrix}, \quad U = \begin{bmatrix} u_0 & u_1 & \cdots & u_{N-1-r} \end{bmatrix}, \tag{27}$$

$$G_d = \begin{bmatrix} g_r & 0 & 0 & \cdots & 0 \\ g_{r+1} & g_r & 0 & \cdots & 0 \\ \vdots & \ddots & \ddots & \ddots & 0 \\ g_{N-1} & g_{N-2} & g_{N-3} & \cdots & g_r \end{bmatrix} \in \Re^{(N-r)\times(N-r)}. \tag{28}$$

It follows that, for sufficiently small $T_S > 0$,

$$\int_0^\infty u^T y ds = T_s U^T Y = T_s U^T G_d U, \quad \int_0^\infty u^T u ds = T_s U^T U. \tag{29}$$

Next, the relationship between the reversed input and output sequences is described. Let $P \in \mathbb{R}^{N\times N}$ be the exchange (or reversal) matrix whose elements are 0 except for anti-diagonal elements being 1's. If $Y = G_d U$, the response under $U' = PU$ is

$$Y' = G_d PU, \tag{30}$$

where

$$y_l' = \sum_{\tau=0}^{l} g_\tau u_{l-\tau}' = \sum_{\tau=0}^{l} g_\tau u_{N-1-l+\tau}. \tag{31}$$

Therefore, defining

$$Y'' = PY', \tag{32}$$

we have

$$y''_l = y'_{N-1-l} = \sum_{\tau=0}^{N-1-l} g_\tau u_{l+\tau} = \sum_{\tau'=l}^{N-1} g_{\tau'-l} u_{\tau'}. \tag{33}$$

The above expression can be rewritten in a compact form as

$$P \underbrace{\left[G_d \overbrace{(PU)}^{u'} \right]}_{Y'} = PG_d PU = G_d^T U. \tag{34}$$

Eqs. (29) and (34) form the input-output relationship that enable machine learning.

3. Distributed algorithms for estimating the eigenvalues of multi-agent and large scale interconnected systems

We start this section by presenting a distributed algorithm for estimating all the eigenvalues in multi-agent systems where the agent is modeled as a single integrator. Next, we discuss how to extend the approach to the case of large scale interconnected systems with unknown system matrix and whose subsystem's dynamics is given by a linear-time invariant system. To this end, distributed data-driven estimation algorithms are presented which only rely on the collected measurements of the system's states (i.e., offline data).

3.1 Model-based distributed algorithm for estimating eigenvalues in multi-agent systems

Consider a multi-agent system consisting of n number of agents and whose overall dynamics is given by

$$\dot{y} = Ay, \tag{35}$$

where $y = \left[y_1, \cdots, y_n\right]^T$ and $y_i \in \mathbb{R}$ denotes the output of the i-th agent. It is assumed that agent i only has access to the local information, that is the i-th row of matrix A, denoted by vector $[A]_{i*}$. Furthermore, the agents can communicate/exchange information with each other via a communication network whose topology is similar to the sparsity of matrix A and is given by a strongly connected directed graph $\mathcal{G}$, that is matrix A is irreducible. For example, matrix A can be a Laplacian or adjacency matrix, as described in Section 2.4. The agents aim to collaboratively estimate all the eigenvalues of A by using only their local information $[A]_{i*}$.

3.1.1 Model-based distributed eigenvalues estimation algorithm

In order to estimate all the eigenvalues of A in a distributed manner, all the agents cooperatively perform the following steps whose detailed analysis can be found in [20]:

1. First, all the agents collaboratively transform matrix A into a nonsingular matrix $\overline{A} = \left[\overline{a}_{ij}\right]$ defined as $\overline{A} = A + cI_n$ with $c \in \mathbb{R}$. Based on Gershgorin theorem, the constant c can be chosen cooperatively by the agents to ensure that $|\overline{a}_{ii}| > \sum_{j \neq i} |\overline{a}_{ij}|$ for all $i = \{1, \cdots, n\}$. To that end, agent i first sets $c_i(0) = \epsilon_i + \sum_j |a_{ij}|$ for arbitrary value of $\epsilon_i > 0$ to ensure $|\overline{a}_{ii}| > \sum_{j \neq i} |\overline{a}_{ij}|$. Next, all the agents should choose a common value c from all the values of $c_i(0)$ so that $|\overline{a}_{ii}| > \sum_{j \neq i} |\overline{a}_{ij}|$ for all $i = \{1, \cdots, n\}$. Specifically, the value of c can be chosen as $c = \max_i c_i(0)$ which can be computed distributively by performing the following maximum consensus protocol [21] for n iterations

$$c_i(k+1) = \max_{j \in \mathcal{N}_i \cup \{i\}} c_j(k), \quad k = 0, 1, \cdots, n. \tag{36}$$

2. After constructing a nonsingular matrix $\overline{A}$ from matrix A, each agent then distributively computes $\overline{A}^{-1}$, denoted by Z, by solving a system of linear equations $\overline{A}Z = I_n$ using its own local information $\left[\overline{A}\right]_{i*}$. To this end, each agent implements the following update rule

$$\hat{Z}_i(k+1) = \hat{Z}_i(k) - \frac{1}{|\mathcal{N}_i|} P_i \left(|\mathcal{N}_i| \hat{Z}_i(k) - \sum_{j \in \mathcal{N}_i} \hat{Z}_j(k) \right), \tag{37}$$

where matrix $\hat{Z}_i(k) \in \mathbb{R}^{n \times n}$ denotes the local estimation of $Z = \overline{A}^{-1}$ by the i-th agent at the k-th iteration and whose initial value is chosen to satisfy $\boxed{A}_{i*}^{T} \hat{Z}_i(0) = [I_n]_{i*}^{T}$. Moreover, matrix $P_i = P_i^T \in \mathbb{R}^{n \times n}$ is an orthogonal projection on the kernel of vector $\boxed{A}_{i*}$, namely $P_i = I_n - \frac{1}{\boxed{A}_{i*}^{T} \boxed{A}_{i*}} \boxed{A}_{i*} \boxed{A}_{i*}^{T}$. Each agent's local estimate $\hat{Z}_i(k)$ under (37) will then converge to $\overline{A}^{-1}$ as $k \to \infty$.

3. Each agent computes all the eigenvalues of A by exploiting the relationship between $\lambda_i(A)$ and $\lambda_i(Z)$ given by $\lambda_i(A) = \frac{1}{\lambda_i(Z)} - c$.

One of the benefits of the above method is that each agent is not only able to distributively estimate all the eigenvalues of A but also the corresponding eigenvectors. Specifically, noting that both matrices A and $\overline{A}^{-1}$ share the same eigenvectors each subsystem can compute distributively the eigenvectors of matrix A from the learned matrix $\overline{A}^{-1}$.

Remark 1. Distributed algorithm (37) converges exponentially to $\overline{A}^{-1}$ [20]. Furthermore, it can be observed from (37) that each agent needs to exchange n^2 values with its neighbors and require to store also n^2 values. One may then ask why each agent does not just flood its row $[A]_{i*}$ to its neighbors so that each agent can then construct matrix A. In contrast to (37), the flooding strategy is not locally adaptable under topology changes [20]. Moreover, convergence of update rule (37) is also guaranteed under time-delay and asynchronous setting [22].

3.1.2 An illustrative example

Consider a multi-agent system consisting of four omnidirectional mobile robots whose kinematic model are given by (13) where y_i denotes its position. Since the robot can move in any directions, its kinematic model can be decoupled for each axis and thus in this example, without loss of generality, we only focus on the motion control of the robots for one axis. The goal is to design control input (velocity) u_i, which depends on the positions of some other robots obtained via a communication network, so that all the robots gather at a common location, e.g., for recharging their batteries. This problem is also known as rendezvous problem [5]. To this end, one can design a consensus protocol given in (14) and assuming the network topology is strongly connected it is ensured that all the four robots gather at a common location. The overall system's closed-loop dynamics can then be written as in (35) where matrix $A = -L$. For example, Laplacian matrix L can be designed as

$$L = \begin{bmatrix} 0.5 & -0.5 & 0 & 0 \\ 0 & 0.4 & 0 & -0.4 \\ -0.8 & -0.2 & 1 & 0 \\ 0 & 0 & -0.8 & 0.8 \end{bmatrix} \tag{38}$$

whose eigenvalues equal to $0,\ 1.2616, 0.7192 \pm 0.5367i$. The second smallest eigenvalue of L measures the network connectivity and convergence rate for reaching consensus [23] while the third smallest eigenvalue provides a metric for ensuring robust connectivity in the presence of single robot failures [24]. The robots can cooperatively estimate the eigenvalues of L using the steps presented in previous subsection. Specifically, the robots set $c = 3$ and initial values $\hat{Z}_i(0)$ as

$$\hat{Z}_1(0) = \begin{bmatrix} 0 & 0 & 0 & 0 \\ -2 & 0 & 0 & 0 \\ 0 & 0 & 0 & 0 \\ 0 & 0 & 0 & 0 \end{bmatrix}, \quad \hat{Z}_2(0) = \begin{bmatrix} 0 & 0 & 0 & 0 \\ 0 & 0 & 0 & 0 \\ 0 & 0 & 0 & 0 \\ 0 & -\frac{5}{2} & 0 & 0 \end{bmatrix},$$
$$\hat{Z}_3(0) = \begin{bmatrix} 0 & 0 & 0 & 0 \\ 0 & 0 & 0 & 0 \\ 0 & 0 & \frac{1}{4} & 0 \\ 0 & 0 & 0 & 0 \end{bmatrix}, \quad \hat{Z}_4(0) = \begin{bmatrix} 0 & 0 & 0 & 0 \\ 0 & 0 & 0 & 0 \\ 0 & 0 & 0 & -\frac{10}{8} \\ 0 & 0 & 0 & 0 \end{bmatrix}. \qquad (39)$$

After executing update rule (37) for 100 iterations, all local estimations $\hat{Z}_i(k)$ converge to $\overline{L}^{-1}$, that is

$$\hat{Z}_i(100) = \begin{bmatrix} 0.2859 & 0.0421 & 0.0009 & 0.0044 \\ 0.0014 & 0.2947 & 0.0062 & 0.0310 \\ 0.0573 & 0.0232 & 0.2505 & 0.0024 \\ 0.0121 & 0.0049 & 0.0527 & 0.2637 \end{bmatrix}, i = \{1,2,3,4\}. \qquad (40)$$

After distributively estimating $\overline{L}^{-1}$, robot j can then calculate the eigenvalues of matrix L given by $\frac{1}{\lambda_i(\hat{Z}_j(100))} - 1$.

3.2 Distributed and data-driven algorithms for estimating eigenvalues of a large scale interconnected system

In this subsection, we extend the setting in the previous subsection to large scale (physically) interconnected systems where its subsystem's dynamics can be modeled as an LTI system, for example power system [25] and thermal model of large buildings [26], which makes the dimension of the overall system very high. It is worth noting that in some applications one may only be interested in estimating the dominant eigenvalues of the overall matrix A. For example, in power systems the dominant eigenvalues, *aka* inter-area oscillation modes, play an important role for wide-area monitoring applications. These slow eigenvalues arise from the oscillations between the coherent areas in power system which may lead to small-signal stability concern and thus needs to be constantly monitored. However, the high dimension of the overall system prevents one from using distributed estimation algorithm presented in the previous subsection. In addition, the large-scale system model (i.e., matrix A) is also often unknown/not available in practice due to geographical constraint or it may change because of perturbation which calls for data-driven methods.

Let us consider a large scale interconnected system divided into r nonoverlapping and coherent clusters where the j-th cluster consists of n_j subsystems. Since the clusters are coherent, one can represent the i-th cluster with an equivalent subsystem whose state $\tilde{x}_i \in \mathbb{R}^p$ is the averaged state of all the subsystems in that cluster, i.e., $\tilde{x}_i = \frac{\sum_{j=1}^{n_i} w_j x_j}{\sum_{j=1}^{n_i} w_j}$ where $w_i > 0$ denotes some weights. Hence, the reduced order model of the large scale interconnected system can be written as

$$\dot{\tilde{x}} = A_r \tilde{x}, \tag{41}$$

where $\tilde{x} = \left[\tilde{x}_1^T, \cdots, \tilde{x}_r^T\right]^T$ and A_r is the reduced system matrix whose eigenvalues correspond to the dominant/slow eigenvalues of the original system matrix A. For details of the modeling, one can refer to [27]. It is assumed that the i-th subsystem has access only to its own sampled state $x_i(k) \triangleq x_i(t)\big|_{t=kT}, k = 0, 1, \ldots$ where T denotes the sampling time. The discrete-time model of (41) is then given by

$$\tilde{x}(k+1) = A_d \tilde{x}(k) \tag{42}$$

where matrix $A_d = e^{A_r T}$. The relation between eigenvalues of both A_r and A_d is given by $\lambda_i(A_d) = e^{\lambda_i(A_r)T}$.

If the eigenvalues of A_r are distinct, one can readily apply the Prony method [14, 28] to estimate the eigenvalues of A_r using the averaged state $\tilde{x}(k)$. In the following, we present an alternative distributed and data-driven method to estimate the eigenvalues of A_r which does not require them to be distinct.

3.2.1 Distributed model learning algorithms

Briefly speaking, the idea is to first learn in a distributed fashion the reduced model A_r from the averaged state $\tilde{x}(k)$ [29]. Without loss of generality and for the sake of simplicity, it is assumed that there exists a virtual agent in each cluster which collects measurement $x_i(k)$ from all subsystems in the i-th cluster to calculate the average state in the corresponding cluster and cooperatively learns the reduced model with the other virtual agents. To that end, we assume that the communication network topology between the virtual agents is given by a strongly connected directed graph.

Assuming that the sampling time is sufficiently small, one can approximate A_d as

$$A_d = I + A_r T \tag{43}$$

whose eigenvalues are given by $\lambda_i(A_d) = 1 + T\lambda_i(A_r)$. Each virtual agent i then collects the following sampled averaged state

$$\begin{aligned} X_i &= [\tilde{x}_i(k_0), \cdots, \tilde{x}_i(k_{m-1})] \in \mathbb{R}^{p \times m}, \\ Y_i &= [\tilde{x}_i(k_0+1), \cdots, \tilde{x}_i(k_{m-1}+1)] \in \mathbb{R}^{p \times m}, \end{aligned} \tag{44}$$

where m denotes the amount of data used for learning. Note that the index $\{k_0, k_1, \cdots, k_{m-1}\}$ does not need to be sequential. Defining matrices $X = \left[X_1^T, \cdots, X_r^T\right]^T$ and $Y = \left[Y_1^T, \cdots, Y_r^T\right]^T$, we have the following relation

$$Y = A_d X. \tag{45}$$

Furthermore, let us set $m = pr$. Given that matrix X is nonsingular, matrix A_d can then be learned by computing

$$A_d = YX^{-1} \tag{46}$$

using only local information available to each virtual agent. To this end, the virtual agents first estimate X^{-1} using update rule similar to (37). Once all the virtual agents compute the estimate of X^{-1}, the i-th virtual agent can then learn the $(p(i-1)+1)-th$ until the pi-th rows of matrix A_d, denoted by $A_{d,i}$, as follows:

$$A_{d,i} = Y_i X^{-1}. \tag{47}$$

Finally, using the learned $A_{d,i}$ the virtual agents can distributively estimate the eigenvalues $\lambda_i(A_d)$ (and $\lambda_i(A_r)$ accordingly) using the method presented in Section 3.1.

One key assumption in the method described above is that all the virtual agents are able to construct matrix X_i from its own state measurement such that the matrix X is nonsingular. When the matrix X is ill-conditions for any given set of sampled measurements, one can alternatively learn matrix A_d by solving a least square problem. To that end, each virtual agent first constructs the measurement matrices X_i, Y_i in (44) with $m > p^2 r$ samples and from (45) satisfy the relation $Y_i = A_{d,i}X$. Next, the agent constructs a vector $a_i \in \mathbb{R}^{p^2 r}$ whose entries equal to the entries of unknown matrix $A_{d,i}$. Equation $Y_i = A_{d,i}X$ can be written as

$$X_i^r a_i = h_i, \tag{48}$$

where matrix $X_i^r \in \mathbb{R}^{pm \times p^2 r}$ and vector $h_i \in \mathbb{R}^{pm}$ are constructed from matrices X and Y_i, respectively. Agent i can then learn the entries of $A_{d,i}$ by solving the following least square problem

$$\hat{a}_i = \arg\min_{a_i} \frac{1}{2} \left\| X_i^r a_i - h_i \right\|_2^2. \tag{49}$$

Remark 2. In order to construct matrix X_i^r and learn the local model, agent i needs to collect X_j from all other agents which requires all-to-all bidirectional communication between the virtual agents. This communication requirement and the size of vector a_i can be reduced if the sparsity structure of matrix A_d is known in advance. After learning the local model, distributed algorithms presented in Section 3.1.1, and whose complexity is discussed in Remark 1, can be adopted to distributively estimate the eigenvalues.

Remark 3. If the structure or property of matrix A_d is known, one can then incorporate this side information as constraints in solving (49).

Remark 4. When the measurements are noisy, one can perform data preprocessing by filtering the noise or smoothing the data before learning the local model.

3.2.2 An illustrative example

Consider an interconnected system of 16 undamped oscillators, divided into 4 coherent clusters as shown in **Figure 1**. Dynamics of the i-th oscillator is given by

$$\Delta\ddot{\delta}_i = -\sum_{j \in \mathcal{N}_i} \frac{\Delta\delta_i - \Delta\delta_j}{r_{ij}}, \tag{50}$$

where $\Delta\delta_i$ denotes the phase angle of oscillator i and r_{ij} represents the reactance of tie lines connecting oscillators i and j. We set r_{ij} of the intra-cluster and intercluster tie lines to be 0.001 per unit and 1 per unit respectively. The reduced order matrix A_r is given by $A_r = \begin{bmatrix} 0 & I \\ L_r & 0 \end{bmatrix}$ where $-L_r$ is a weighted Laplacian matrix and the state $\tilde{x} = \left[\Delta\delta_{r,1}, \cdots, \Delta\delta_{r,4}, \Delta\dot{\delta}_{r,1}, \cdots, \Delta\dot{\delta}_{r,4}\right]^T$ with $\Delta\delta_{r,i}$ denotes the average phase angle of the equivalent oscillator in the i-th cluster. Using the slow-fast time-scale separation principle [30], matrix L_r can be analytically calculated as

$$L_r = \begin{bmatrix} -0.4997 & 0.2498 & 0.0001 & 0.2498 \\ 0.2498 & -0.4997 & 0.2498 & 0.0001 \\ 0.0001 & 0.2498 & -0.4997 & 0.2498 \\ 0.2498 & 0.0001 & 0.2498 & -0.4997 \end{bmatrix}. \tag{51}$$

In order to learn distributively matrix A_r, a virtual agent is assigned to each cluster which can communicate with each other as shown in **Figure 1**. Each virtual agent computes the average state for each area as illustrated in **Figure 2a**. Each virtual agent then distributively learn the corresponding rows of A_r by solving (49) where the number of data $m = 1000$ and the index $k_{j+1} - k_j = 10$ in (44) for all j, see **Figure 2b**. Note that the sampling time equals to $T = 0.001s$. In addition, the virtual agent also incorporates the side information regarding the sparsity structure of A_r and the property of the Laplacian $-L_r$, discussed in Section 2.4, when solving its least square problem. The learned model is given by

$$\hat{L}_r = \begin{bmatrix} -0.5004 & 0.2487 & 0.0014 & 0.2503 \\ 0.2496 & -0.5001 & 0.2497 & 0.0008 \\ 0.0007 & 0.2504 & -0.5006 & 0.2496 \\ 0.2507 & 0.0016 & 0.2486 & -0.5009 \end{bmatrix}. \tag{52}$$

The comparison between the data and the trajectories of the reduced dynamical system using the learned model is shown in **Figure 2b**. It can be observed that the virtual agents are able to learn accurately the reduced order model A_r. Finally, the

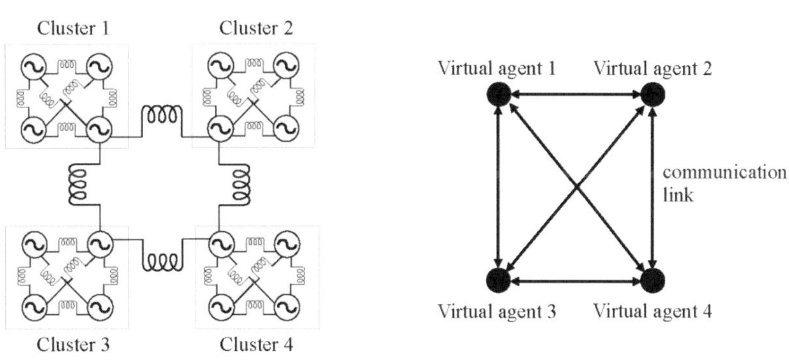

Figure 1.
Left: 4-cluster 16 second-order oscillators. Right: virtual agents representing each cluster and their communication network topology.

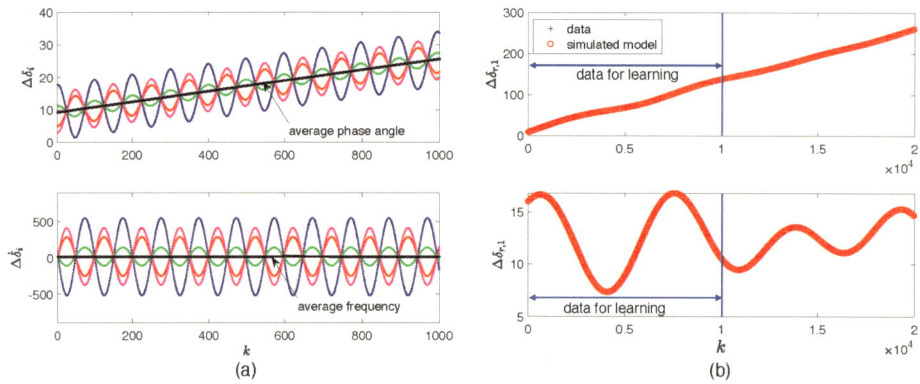

Figure 2.
(a) Snapshots of true states and average states of oscillators in cluster 1; (b) Comparison between data and trajectories generated using the learned reduced order model in cluster 1.

comparison between eigenvalues of the analytically calculated matrix A_r and the eigenvalues of the learned matrix $\hat{A}_r$ is given by

$$\begin{aligned} \text{true eigenvalues}: &\quad 0, 0, \pm 0.7070i, \pm 0.7070i, \pm 0.9996i, \\ \text{estimated eigenvalues}: &\quad 0, 0, \pm 0.7074i, \pm 0.7091i, \pm 0.9994i. \end{aligned}$$

4. Machine learning for passive and passivity-short systems

In this section, two methods are explored to determine the indices ϵ and ρ of passive and passivity-short systems. These indices can be found using the eigenvalue-based approach of linear matrix inequality (LMI) and the data-driven technique proposed in [19, 31] for passive systems. The LMI approach requires perfect knowledge of the system but offers the highest level of precision in determining the passivity indices, as it avoids the errors associated with discrete-time approximations, such as those introduced by sampling and discretization. In comparison, the data-driven approach is less precise but offers an approach of finding passivity or passivity short indices solely and directly from input output data, providing the alternative for unknown systems and avoiding the step of model identification.

4.1 Direct method: Linear matrix inequality (LMI) approach

The LMI approach is an eigenvalue based method that leverages the continuous time system for defining an inequality condition that expresses constraints on a system using matrices, where the passivity constraint is integrated directly into the condition. By formulating the passivity conditions as LMIs, convex optimization techniques can be utilized to solve for the indices.

Consider system (1) and the Lyapunov function $V = 0.5x^T Sx$, where S is a positive definite matrix. The time derivative of Lyapunov function is given by

$$\dot{V} = \frac{1}{2} x^T (SA + A^T S)x + x^T SBu, \tag{53}$$

which satisfy the following dissipativity condition

$$\dot{V} \leq u^T y - \frac{\epsilon}{2}\|u\|^2 - \frac{\rho}{2}\|y\|^2, \tag{54}$$

if and only if the following matrix inequality (i.e., in terms of its minimum eigenvalue being non-negative) hold:

$$W(\epsilon,\rho) = \begin{bmatrix} -A^T S - SA - \rho C^T C & -SB + C^T - \rho C^T D \\ -B^T S + C - \rho D^T C & -\epsilon I + D + D^T - \rho D^T D \end{bmatrix} \geq 0. \tag{55}$$

Leveraging semi-definite program solvers for optimization, passivity indices ϵ and ρ can be determined by following the procedure defined in Algorithm 1.

Algorithm 1 Passivity Indices Using LMI

Solve
$$\tilde{\epsilon} = \arg\max_S \epsilon \quad \text{subject to} \quad W(\epsilon, 0) \geq 0, \quad S > 0$$
if $\tilde{\epsilon} > 0$ **then**
  The system is passive
  The ISP index $\epsilon^* = \tilde{\epsilon}$
  The OSP index ρ^* is determined by
$$\rho^* = \arg\max_S \rho, \quad \text{subject to} \quad W(0,\rho) \geq 0, \quad S > 0$$
  The IOSP indices ρ (or ϵ) can be found for any fixed value of ϵ (or ρ) less than its upper bound using
$$\epsilon^* (\text{or } \rho^*) = \arg\max_S \epsilon \ (\text{or } \rho^*), \quad \text{subject to} \quad W(\epsilon,\rho) \geq 0, \quad S > 0$$
end if
if $\tilde{\epsilon} < 0$ **then**
  The system is passivity short
  Set $\epsilon = -1$ to admit unique solutions
  The OFPS index ρ^* is determined by
$$\rho^* = \arg\max_S \rho, \quad \text{subject to} \quad W(-1,\rho) \geq 0, \quad S > 0$$
  The IFPS index ϵ is determined, for any $\rho \in [0, \rho^*)$, using
$$\epsilon =_S \epsilon, \quad \text{subject to} \quad W(\epsilon,\rho) \geq 0, \quad S > 0$$
end if

Remark 5. The above LMI solution is computationally efficient as it has the complexity of linear programming. Should the system dynamics are linear but contain parameterizable and bounded uncertainties, one could adopt the model of so-called interval systems. In this case, the matrices $\{A, B, C, D\}$ may have their known parts and their uncertain parts. For example, system matrix A becomes $A + \Delta A$, where entries of ΔA belong to certain known intervals. In such cases, the above LMI solution can be extended to these interval systems, and the reader is referred to relevant literature.

4.2 Indirect method: Data-driven approach

In previous sections, analytical methods for determining passivity indices, using model-based approaches, were examined. While these methods provide precise methods for determining passivity, they are explicitly dependent on the mathematical model of the system. However, in practical applications, the system model cannot be obtained or determined due to the complex nature of the system or unknown knowledge of parameters. To address these challenges, we extend the data-driven approaches in [19, 31] to find passivity and passivity short indices on both the input and output channels of a model using data only.

Assume that system (1) can be tested but its model is unknown. Then, its output data can be collected upon feeding an input into the system. If $x(0) = 0, x(\infty) = 0$, we have

$$0 \leq \int_0^\infty u^T y ds - \frac{\epsilon}{2}\int_0^\infty \|u\|^2 ds - \frac{\rho}{2}\int_0^\infty \|y\|^2 ds \tag{56}$$

where ϵ, $\rho > 0$ are the largest possible values to satisfy the above inequality. It can be seen that, if ρ (or ϵ) is given then

$$\epsilon \leq \frac{\int_0^\infty (u^T y + y^T u - \rho\|y\|^2) ds}{\int_0^\infty \|u\|^2 ds} \quad \text{or} \quad \rho \leq \frac{\int_0^\infty (u^T y + y^T u - \epsilon\|u\|^2) ds}{\int_0^\infty \|y\|^2 ds} \tag{57}$$

where ϵ (or ρ) > 0 is the largest value satisfying the above inequality. This can be further refined into objective functions

$$\epsilon = \max_U f_\epsilon(U) \quad \text{or} \quad \rho = \max_U f_\rho(U) \tag{58}$$

where

$$f_\epsilon(U) = \frac{U^T Y + Y^T U - \rho Y^T Y}{U^T U} \quad \text{or} \quad f_\rho(U) = \frac{U^T Y + Y^T U - \epsilon U^T U}{Y^T Y}. \tag{59}$$

To solve Eq. (58), a gradient descent method can be implemented using the update rule

$$U^{(k+1)} = U^{(k)} - \delta^{(k)} \nabla f_z\left(U^{(k)}\right) \tag{60}$$

where $z \in [\rho, \epsilon]$, the optimal step size $\delta^{(k)}$ is given by Eq. (67) and $\nabla f_z\left(U^{(k)}\right)$ is the gradient of f_z defined in Algorithm 3 of the Appendix.

Algorithm 2 provides the procedure to determine passivity indices ϵ and ρ in a similar manner to Algorithm 1, and Algorithm 3 defines the gradient descent function used in Algorithm 2.

Remark 6. The above data-driven approach is iterative and computationally simple, and its convergence property depends upon the specific numerical search algorithm used. In the following examples, the standard gradient search algorithm is used, and its convergence is optimized by online implementing the optimal stepsize derived in the Appendix.

4.3 Illustrative examples

In this section, we demonstrate the effectiveness of the proposed algorithms on simple second order systems and a real world synchronous generator system of order 6.

Algorithm 2 Passivity Indices Using Data Driven Indirect Approach

1: Solve the following where $\rho = 0$

$$\tilde{\epsilon} = PassivityGradientDescent\left(f_\epsilon, \nabla f_\epsilon, U_\epsilon^{(0)}, \delta_\epsilon, K, 0\right)$$

2: **if** $\tilde{\epsilon} > 0$ **then**

3: **The system is passive**

4: The ISP index $\epsilon^* = \tilde{\epsilon}$

5: The OSP index ρ^* is determined by

$$\rho^* = PassivityGradientDescent\left(f_\rho, \nabla f_\rho, U_\rho^{(0)}, \delta_\rho, K, 0\right)$$

6: The IOSP indices ρ (or ϵ) can be found for any fixed value of ϵ (or ρ) less than its upper bound using

$$\epsilon^* = PassivityGradientDescent\left(f_\epsilon, \nabla f_\epsilon, U_\epsilon^{(0)}, \delta_\epsilon, K, \rho\right)$$

$$\rho^* = PassivityGradientDescent\left(f_\rho, \nabla f_\rho, U_\rho^{(0)}, \delta_\rho, K, \epsilon\right)$$

7: **end if**

8: **if** $\tilde{\epsilon} < 0$ **then**

9: **The system is passivity short**

10: Set $\epsilon = -1$ to admit unique solutions

11: The OFPS index ρ^* is determined by

$$\rho^* = PassivityGradientDescent\left(f_\rho, \nabla f_\rho, U_\rho^{(0)}, \delta_\rho, K, \epsilon\right)$$

12: The IFPS index ? is determined, for any $\rho \in [0, \rho^*)$, using

$$\epsilon^* = PassivityGradientDescent\left(f_\epsilon, \nabla f_\epsilon, U_\epsilon^{(0)}, \delta_\epsilon, K, \rho\right)$$

13: **end if**

4.3.1 Simple 2nd order systems

Consider the following two input-output stable, continuous time systems:

$$\dot{x} = \begin{bmatrix} 0 & 1 \\ -2 & -3 \end{bmatrix} x + \begin{bmatrix} 0 \\ 1 \end{bmatrix} u \quad y = \begin{bmatrix} 1 & 0 \end{bmatrix} x + 0.5u. \tag{61}$$

$$\dot{x} = \begin{bmatrix} 0 & 1 \\ -2 & -3 \end{bmatrix} x + \begin{bmatrix} 0 \\ 1 \end{bmatrix} u \quad y = \begin{bmatrix} 1 & 0 \end{bmatrix} x. \tag{62}$$

These systems are discretized using a first-order hold with a sampling rate of 0.01 and $T_s = 2500$. The initial inputs were chosen to be

$$U_\epsilon^{(0)}(t) = \frac{\sin(0.2\pi t)}{\|\sin(0.2\pi t)\|}, \quad U_\rho^{(0)}(t) = \frac{\sin(2\pi t)}{\|\sin(2\pi t)\|} \tag{63}$$

Figure 3 presents the passivity index results for passive system (61), and **Figure 4** shows the results for passivity short system (62). In both figures, the Nyquist diagrams provide visual confirmation of the systems' stability. The convergence of the ISP and IFPS indices, ϵ, is depicted in the top-right plot, illustrating how, over the prescribed number of iterations k, the estimate of ϵ from the indirect approach converges to the true value of ϵ given by the direct approach. Similarly, the convergence of OSP and OFPS indices ρ are shown in the bottom-left plots. Finally, the bottom-right plots present the Pareto fronts for IOSP and IOPS indices. The Pareto front represents the trade-offs between ϵ and ρ, where optimality is achieved by minimizing one parameter while solving for the other. For passive systems, the feasible region is finite and constrained to the first quadrant, indicating that both ϵ and ρ

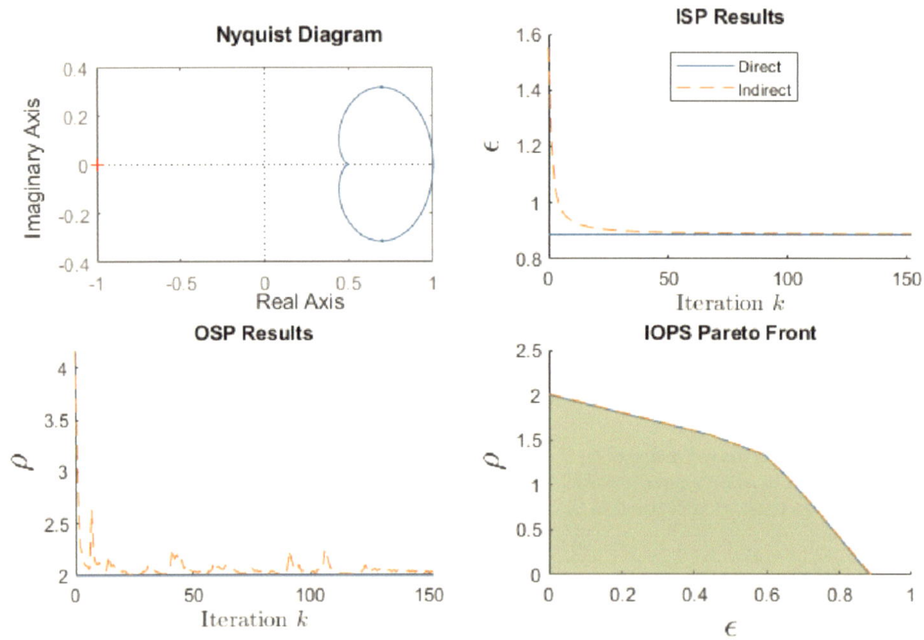

Figure 3.
Passivity results for system (61). Nyquist diagram (top left)*, convergence of indirect ϵ to direct ϵ* (top right)*, convergence of indirect ρ to direct ρ* (bottom left)*, pareto front* (bottom right)*.*

remain non-negative. In contrast, the passivity-short system exhibits an unbounded region, with ϵ extending to negative infinity, reflecting the greater flexibility and reduced constraints inherent in passivity-short conditions.

4.3.2 A real-world example

Consider the following 6th order model of a turbine generator [32], in form of system (1) with matrices:

$$A = \begin{bmatrix} 0 & 377 & 0 & 0 & 0 & 0 \\ -0.2673 & 0 & -0.2946 & 0 & 0 & 0.211 \\ -0.2763 & 0 & -0.580 & 0.1695 & 0 & 0 \\ -66.3405 & 0 & -535.1923 & -20 & 0 & 0 \\ 0 & -0.09 & 0 & 0 & -3.333 & 0 \\ 0 & 0 & 0 & 0 & 1.0 & -1.0 \end{bmatrix},$$

$$B = \begin{bmatrix} 0 & 0 & 0 & 0 & 3.3333 & 0 \end{bmatrix}^T$$
$$C = \begin{bmatrix} 1.2668 & 0 & 1.3966 & 0 & 0 & 0 \end{bmatrix},$$
$$D = [0].$$

Compared to that in [32], the above is a reduced order model in which the excitation control dynamics are removed. The discretized system is obtained using a least squares approximation with a sampling rate of 0.01 and $T_s = 2500$ and it can be

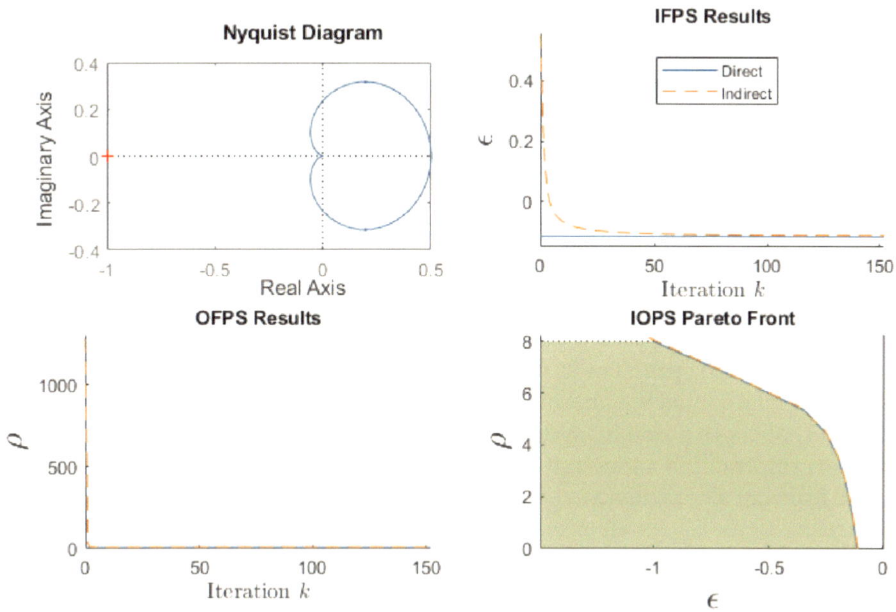

Figure 4.
*Passivity results for system (62). Nyquist diagram (*top left*), convergence of indirect ϵ to direct ϵ (*top right*), convergence of indirect ρ to direct ρ (*bottom left*), pareto front (*bottom right*).*

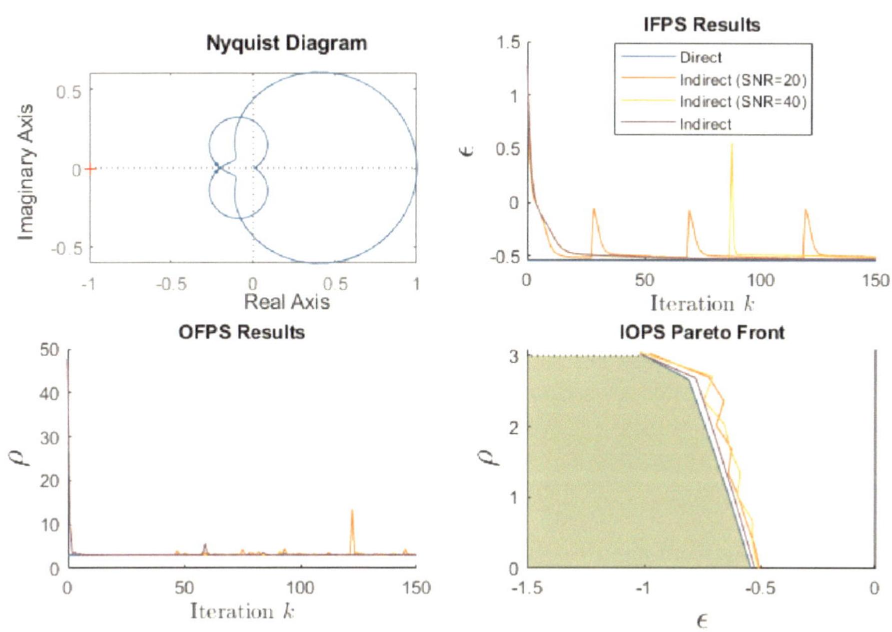

Figure 5.
*Passivity results for turbine generator. Nyquist diagram (*top left*), convergence of indirect ϵ, with and without noise, to direct ϵ (*top right*), convergence of indirect ρ to direct ρ with and without noise (*bottom left*), pareto front with and without noise (*bottom right*).*

shown that the system is passivity short. The initial inputs from (63) are reused, and to demonstrate practicality, the indirect approach was solved with and without the presence of noise. A white Gaussian noise, with a specified signal to noise ration (SNR) given in dB, was applied to the outputs of the system that are determined directly from U. **Figure 5** provides the results for the turbine generator system when both direct and indirect passivity approaches applied as well as with and without noises.

5. Conclusion

In this chapter the problem of determining eigenvalues, and thereby determining stability, for interconnected and multi-agent dynamic systems is investigated. To overcome challenges posed by system size, complexity, topology changes, and incomplete information about subsystems, scalable, data driven estimation and machine learning algorithms are proposed. First, two distributed algorithms capable of estimating all the eigenvalues of multi-agent cooperative systems and/or large-scale interconnected systems, including dominant eigenvalues, was proposed. Additionally, the input-output stability of subsystems is investigated and the extension of eigen-analysis is made to study passivity constraints using system matrices or purely input output data. This analysis was further advanced by leveraging machine learning algorithms to learn the passivity properties without prior knowledge of the system dynamics. The effectiveness of the proposed algorithms was demonstrated through numerical examples. The work presented in this chapter contributes to laying a more concrete foundation for scalable stability analysis, enabling the use of data-driven and machine learning algorithms in complex interconnected systems.

Open source code availability

The repository containing the implemented algorithms presented in this chapter can be acquired at https://forms.gle/z3T75RMRrWM7GSfb9.

Acknowledgements

The work of Zhihua Qu was supported in part by the U.S. Department of Energy's under Award DE-EE0007998, Award DE-EE0009028, Award DEEE0009152, Award DE-EE0009339, and Award DE-AC05-76RL01830. The work of Azwirman Gusrialdi was supported by the Research Council of Finland under Academy Project 330073.

A. Appendix

The gradient descent algorithm to determine the passivity index, as well as the derivation for the optimal step size of the indirect approach is provided.

A.1 Gradient descent for passivity

To minimize the optimization functions (58), we employ the well-known strategy of gradient descent. This method involves iteratively updating U by moving in the direction opposite to the gradient of the function f at each step, thereby approaching the minimum of f. Specifically. the passivity gradient descent approach is provided in Algorithm 3 where an input U is chosen initially, then for k iterations, various projections of the input are passed to the system to calculate $f(U)$, its gradient $\nabla f(U)$, and the optimal stepsize δ. At the end of each iteration, $\nabla f(U)$, and δ are used to update the input U by Eq. (60).

A.2 Optimal Stepsize δ

To increase the rate of convergence of estimation, the stepsize $\delta^{(k)}$ can be updated each iteration. Let f_z, where $z \in [\epsilon, \rho]$ be generalized to f, then the optimal step size is given by

$$\delta^{(k)} = \arg\min_{\delta \in \mathbb{R}} f\left(U^{(k)} - \delta^{(k)} \nabla f\left(U^{(k)}\right)\right). \tag{64}$$

Algorithm 3 Passivity Gradient Descent

Choose an initial control input $U^{(0)} \in \mathbb{R}^{1\times N}$
for k K **do**
  Simulate to get $Y_1^{(k)}, Y_2^{(k)}$, and $Y_3^{(k)}$, under inputs $U^{(k)}, PU^{(k)}$ and $PY_1^{(k)}$.
  if ρ is given **then**
    Calculate $\epsilon^{(k)} = f_\epsilon(U^k)$
    Calculate $\nabla f\left(U^{(k)}\right) = \nabla f_\epsilon^{(k)} = \frac{2(Y_1 + PY_2 - \rho PY_3)}{U^T U} - 2f_\epsilon(U)\frac{U}{U^T U}$
    Simulate to get $Y_4^{(k)}, Y_5^{(k)}$, and $Y_6^{(k)}$ under input $\nabla f_\epsilon^{(k)}, P\nabla f_\epsilon^{(k)}$ and $PY_4^{(k)}$
    Calculate the step size $\delta^{(k)}$ by Eq. (67) with

$$M_1 = \begin{bmatrix} \left(U^{(k)}\right)^T\left(Y_1^{(k)} + PY_2^{(k)} - \rho PY_3^{(k)}\right) & -\left(U^{(k)}\right)^T\left(Y_4^{(k)} + PY_5^{(k)} - \rho PY_6^{(k)}\right) \\ -\left(\nabla f_\epsilon^{(k)}\right)^T\left(Y_1^{(k)} + PY_2^{(k)} - \rho PY_3^{(k)}\right) & \left(\nabla f_\epsilon^{(k)}\right)^T\left(Y_4^{(k)} + PY_5^{(k)} - \rho PY_6^{(k)}\right) \end{bmatrix}$$

$$M_2 = \begin{bmatrix} \left(U^{(k)}\right)^T U^{(k)} & -\left(U^{(k)}\right)^T \nabla f_\epsilon^{(k)} \\ -\left(\nabla f_\epsilon^{(k)}\right)^T U^{(k)} & \left(\nabla f_\epsilon^{(k)}\right)^T \nabla f_\epsilon^{(k)} \end{bmatrix}$$

    Update $U^{(k+1)} = U^{(k)} - \delta^{(k)} \nabla f_\epsilon^{(k)}$
  else if ϵ is given **then**
    Calculate $\rho^{(k)} = f_\rho(U^k)$
    Calculate $\nabla f_\rho\left(U^{(k)}\right) = \nabla f_\rho^{(k)} = \frac{2(Y_1 + PY_2 - \epsilon U)}{Y_1^T Y_1} - 2f_\rho(U)\frac{Y_3}{Y_1^T Y_1}$
    Simulate to get $Y_4^{(k)}, Y_5^{(k)}$, and $Y_6^{(k)}$ under input $\nabla f_\rho^{(k)}, P\nabla f_\rho^{(k)}$ and $PY_4^{(k)}$
    Calculate the step size $\delta^{(k)}$ by Eq. (67) with

$$M_1 = \begin{bmatrix} \left(U^{(k)}\right)^T\left(Y_1^{(k)} + PY_2^{(k)} - \epsilon U^{(k)}\right) & -\left(U^{(k)}\right)^T\left(Y_4^{(k)} + PY_5^{(k)} - \epsilon \nabla f_\rho^{(k)}\right) \\ -\left(\nabla f_\rho^{(k)}\right)^T\left(Y_1^{(k)} + PY_2^{(k)} - \epsilon U^{(k)}\right) & \left(\nabla f_\rho^{(k)}\right)^T\left(Y_4^{(k)} + PY_5^{(k)} - \epsilon \nabla f_\rho^{(k)}\right) \end{bmatrix}$$

$$M_2 = \begin{bmatrix} \left(U^{(k)}\right)^T PY_3 & -\left(U^{(k)}\right)^T PY_6 \\ -\left(\nabla f_\rho^{(k)}\right)^T PY_3 & \left(\nabla f_\rho^{(k)}\right)^T PY_6 \end{bmatrix}$$

    Update $U^{(k+1)} = U^{(k)} - \delta^{(k)} \nabla f_\rho^{(k)}$
  end if
  Run simulation to get $Y_1^{(k+1)}$ under input $U^{(k+1)}$
end for

It follows from Eq. (59) that

$$f(U - \delta \nabla f(U)) = \frac{[1 \quad \delta]M_1[1 \quad \delta]^T}{[1 \quad \delta]M_2[1 \quad \delta]^T} = \frac{M_{1,11} + 2M_{1,12}\delta + M_{1,22}\delta^2}{M_{2,11} + 2M_{2,12}\delta + M_{2,22}\delta^2} \tag{65}$$

and that

$$\begin{aligned}
&\frac{1}{2}\left(M_{2,11} + 2M_{2,12}\delta + M_{2,22}\delta^2\right)^2 \frac{\partial f}{\partial \delta} \\
&= (M_{1,12}M_{2,11} - M_{2,12}M_{1,11}) + (M_{1,22}M_{2,11} - M_{2,22}M_{1,11})\delta \\
&+ (M_{1,22}M_{2,12} - M_{2,22}M_{1,12})\delta^2 \\
&\triangleq c + b\delta + a\delta^2,
\end{aligned} \tag{66}$$

from which optimal value δ^* can be solved by setting the above expression equal to zero. That is,

$$\delta^* = \frac{-b \pm \sqrt{b^2 - 4ac}}{2a}, \tag{67}$$

in which the (smaller) positive solution should be chosen.

Author details

Kenneth McDonald[1], Zhihua Qu[1*] and Azwirman Gusrialdi[2]

1 University of Central Florida, Orlando, United States

2 Tampere University, Tampere, Finland

*Address all correspondence to: qu@ucf.edu

References

[1] Kailath T. Linear Systems. Englewood Cliffs, NJ: Prentice Hall; 1980

[2] Ogata K. Modern Control Engineering. Englewood Cliffs, NJ: Prentice Hall; 1990

[3] Khalil HK. Nonlinear Systems. Upper Saddle River, NJ: Prentice Hall; 2002

[4] Willems JC. Dissipative dynamical systems part i: General theory. Archive for Rational Mechanics and Analysis. 1972;**45**:321-351. DOI: 10.1007/BF00276493

[5] Qu Z. Cooperative Control of Dynamical Systems. London: Springer Verlag; 2009

[6] Zhihua Q, Simaan MA. Modularized design for cooperative control and plug-and-play operation of networked heterogeneous systems. Automatica. 2014;**50**(9):2405-2414

[7] Poonawala HA, Spong MW. Decentralized estimation of the algebraic connectivity for strongly connected networks. In: American Control Conference. Chicago, IL, USA: IEEE; 2015. pp. 4068-4073

[8] Gusrialdi A, Zhihua Q, Hirche S. Distributed link removal using local estimation of network topology. IEEE Transactions on Network Science and Engineering. 2019;**6**(3):280-292

[9] Alizadeh R, Bijani S, Shakeri F. Distributed consensus-based estimation of the leading eigenvalue of a non-negative irreducible matrix. Parallel Computing. 2024;**122**:103113. DOI: 10.1016/j.parco.2024.103113

[10] Charalambous T, Rabbat MG, Johansson M, Hadjicostis CN. Distributed finite-time computation of digraph parameters: Left-eigenvector, out-degree and spectrum. IEEE Transactions on Control of Network Systems. 2015;**3**(2):137-148

[11] Fernández OD, Tiistola S, Gusrialdi A. Real-time data-driven electromechanical oscillation monitoring using dynamic mode decomposition with sliding window. In: 11th IFAC Symposium on Control of Power and Energy Systems. IFAC-PapersOnLine; 2022. pp. 158-163

[12] Fernández OD, Iqbal M, Gusrialdi A. An improved dynamic mode decomposition for real-time electromechanical oscillation monitoring in power systems: The impact of ultra-low frequency modes and its removal strategy. IET Generation Transmission and Distribution. 2023;**17**(20):4574-4591

[13] Gusrialdi A, Zhihua Q. Distributed data-driven power iteration for strongly connected networks. In: European Control Conference. Delft, Netherlands: IEEE; 2021. pp. 87-92

[14] Khazaei J, Fan L, Jiang W, Manjure D. Distributed prony analysis for real-world pmu data. Electric Power Systems Research. 2016;**133**:113-120

[15] Deplano D, Congiu C, Giua A, Franceschelli M. Distributed estimation of the laplacian spectrum via wave equation and distributed optimization. In: 22nd IFAC World Congress. Yokohama, Japan: IFAC-PapersOnLine; 2023. pp. 6952-6957

[16] Hayhoe M, Barreras F, Preciado VM. A dynamical approach to efficient eigenvalue estimation in general multiagent networks. Automatica. 2022; **140**:110234

[17] Venkateswaran DB, Qu Z. Passivity-short bilateral teleoperation with

communication delays. In: IEEE International Conference on Systems, Man, and Cybernetics. Miyazaki, Japan: IEEE; 2018. pp. 1275-1281

[18] Joo Y, Harvey R, Zhihua Q. Preserving and achieving passivity-short property through discretization. IEEE Transactions on Automatic Control. 2020;**65**(10):4265-4272. DOI: 10.1109/TAC.2019.2954361

[19] Tanemura M, Azuma S-i. Efficient data-driven estimation of passivity properties. IEEE Control Systems Letters. 2019;**3**(2):398-403. DOI: 10.1109/LCSYS.2018.2887241

[20] Gusrialdi A, Zhihua Q. Distributed estimation of all the eigenvalues and eigenvectors of matrices associated with strongly connected digraphs. IEEE Control Systems Letters. 2017;**1**(2):328-333

[21] Nejad BM, Attia SA, Raisch J. Max-consensus in a max-plus algebraic setting: The case of fixed communication topologies. In: International Symposium on Information, Communication and Automation Technologies. Sarajevo, Bosnia and Herzegovina: IEEE; 2009. pp. 1-7

[22] Liu J, Mou S, Stephen Morse A. Asynchronous distributed algorithms for solving linear algebraic equations. IEEE Transactions on Automatic Control. 2018;**63**(2):372-385

[23] Asadi MM, Khosravi M, Aghdam AG, Blouin S. Generalized algebraic connectivity for asymmetric networks. In: American Control Conference. Boston, MA, USA: IEEE; 2016. pp. 5531-5536

[24] Zareh M, Sabattini L, Secchi C. Distributed Laplacian Eigenvalue and Eigenvector Estimation in Multi-Robot Systems. Cham: Springer International Publishing; 2018. pp. 191-204

[25] Liu J, Gusrialdi A, Hirche S, Monti A. Joint controller-communication topology design for distributed wide-area damping control of power systems. In: 18th IFAC World Congress. Milano, Italy: IFAC-PapersOnLine; 2011. pp. 519-525

[26] Moroşan P-D, Bourdais R, Dumur D, Buisson J. Building temperature regulation using a distributed model predictive control. Energy and Buildings. 2010;**42**(9):1445-1452

[27] Chow JH. Power System Coherency and Model Reduction. Vol. 84. New York, NY: Springer; 2013

[28] Gusrialdi A, Qu Z. Data-driven distributed algorithms for estimating eigenvalues and eigenvectors of interconnected dynamical systems. In: Proceedings of 21st IFAC World Congress. Berlin, Germany: IFAC-PapersOnLine; 2020. pp. 52-57

[29] Gusrialdi A, Chakrabortty A, Zhihua Q. Distributed learning of mode shapes in power system models. In: Proceedings of IEEE Conference on Decision and Control. Miami, FL: IEEE; 2018. pp. 4002-4007

[30] Chow J, Kokotovic P. Time scale modeling of sparse dynamic networks. IEEE Transactions on Automatic Control. 1985;**30**(8):714-722

[31] Koch A, Montenbruck JM, Allgöwer F. Sampling strategies for data-driven inference of input–output system properties. IEEE Transactions on Automatic Control. 2021;**66**(3):1144-1159. DOI: 10.1109/TAC.2020.2994894

[32] Smaili YA, Alouani AT. An H/sub infinity / governor exciter controller design for a power system. In: [Proceedings 1992] The First IEEE Conference on Control Applications. Vol. 2. Dayton, OH, USA: IEEE; 1992. pp. 770-775. DOI: 10.1109/CCA.1992.269750

Chapter 7

Multivariate Linear Model for Data Analysis and Machine Learning and the Theory and Practice of Eigenvalues in Mitigating Multicollinearity

Tor A. Kwembe

Abstract

The chapter introduces a multivariate high dimensional linear model for large dataset analytics and machine learning and the mathematical derivation of its parameters. We covered regression techniques and analysis for multidimensional datasets, mitigating multicollinearity, and dimension reduction techniques and the decision tree classifier method that is applied to Machine Learning and Artificial Intelligence. We further explained collinearity and multicollinearity in a matrix perspective approach and mitigation methods to improve machine learning and data analytics algorithms and techniques. We demonstrated with proofs that when an eigenvalue of a dataset is zero or very near zero, collinearity or multicollinearity exists among the features of the dataset. We also showed that Principal Component Analysis (PCA) is a method for mitigating multicollinearity among a list of several other methods. The chapter covers the Principal Component Analysis (PCA) method for high dimension data reduction and feature selection in detail, and introduced an example of its applications to a network intrusion detection system data to illustrate the theory and practice of eigenvalues and eigenvectors in modern engineering.

Keywords: data analysis, multivariate linear models, machine learning, eigenvalues and eigenvectors, mitigating multicollinearity, principal component analysis, network intrusion

1. Introduction

In these notes, we have introduced a multivariate linear model for large data analytics and supervised machine learning and illustrate the role eigenvalues play in identifying and mitigating multicollinearity. Multivariate or high dimensional datasets are common in the era of big data analytics and electronic data storage. Multivariate data present many challenges for statistical visualization, analysis, and modeling [1–3]. The inherent difficulties in multivariate linear models for predictive data

IntechOpen

analytics and regression are that of visualizing data that has many variables and the existence of multicollinearity in machine learning algorithms and regression analysis. Because of these challenges, multivariate data analytics and machine learning methods often begin with some type of dimension simplification and reduction to approximate data in lower dimensional space. Principal component analysis (PCA) method is a rigorous computational method for dataset dimension simplification and reduction. It is a statistical technique that has profound applications in many real-world modern engineering fields such as face recognition, image compression, network intrusion detection systems, cybersecurity, and is a common technique for gaining insights into patterns in high dimension datasets. In general, the PCA method generates new set of features or variables, called principal components ordered according to the size of the eigenvalues of the normalized matrix of objects and features. Each principal component can be expressed as a linear combination of the original features or variables. They are orthogonal to each other or orthonormal in the case of a normalized matrix of objects and features, so there is no redundant information. They form an orthogonal/orthonormal basis for the space of the data.

In Section 2, we presented a generalized multivariate/high-dimensional linear model that is common for data analytics and supervised machine learning algorithms. We have given a step-by-step method of optimizing the model parameters using the method of Ordinary Least Squares (OLS) estimate which is known to be unbiased. We identify in the model why collinearity may exist and the methods for identifying multicollinearity, visually and analytically. In Section 3, we considered the matrix form of the high dimensional linear model and stated conditions for the existence of multicollinearity in terms of the eigenvalues of the covariance matrix of the objects and features of the dataset. In Section 4, we used the methods of eigenvalues and eigenvectors in Section 3 to illustrate the Principal Component Analysis method of mitigating multicollinearity in high dimension datasets. In Section 5, we show how to use Principal Components Analysis to fit a multilinear linear regression in Network intrusion detection systems using the high-dimensional CICIDS2017 network intrusion dataset which contains benign and the current and frequently occurring cyberattacks with multicollinear features [4]. In this chapter, we have only used the PCA method for mitigating multicollinearity and high dimension data reduction rather than the OLS driven methods of Ridge, LASSO, and the generalized ELASTIC Net.

2. Multivariate linear models for data analysis and machine learning

In this section, we will introduce a higher dimensional linear model of features of a dataset of the form

$$\mathbf{y} = \mathbf{X}\boldsymbol{\beta} + \boldsymbol{\varepsilon}, \tag{1}$$

where.

$\mathbf{y} = (y_1, y_2, \dots, y_n)^T = \begin{bmatrix} y_1 \\ \vdots \\ y_n \end{bmatrix}$, $\mathbf{X} = (X_0, X_1, X_2, \dots, X_P)$ is a vector whose components are the column vectors of features/variables and the constant $X_0 = \begin{bmatrix} 1 \\ \vdots \\ 1 \end{bmatrix}$, and $X_i = (x_{1,i}, x_{2,i}, \dots, x_{n,i})^T, i = 1, 2, \dots, P$, are the features in a given dataset and $\boldsymbol{\beta}$, the coefficients/parameters to be determined are given as $\boldsymbol{\beta} = (\beta_0, \beta_1, \beta_2, \dots, \beta_P)^T = \begin{bmatrix} \beta_0 \\ \vdots \\ \beta_P \end{bmatrix}$, and $\boldsymbol{\varepsilon}$, the residuals or errors vector is given as $\boldsymbol{\varepsilon} = (\varepsilon_1, \varepsilon_2, \dots, \varepsilon_n)^T$. Thus, $\mathbf{X}$ is a cleaned dataset in matrix form given as:

$$\mathbf{X} = \begin{bmatrix} 1 & x_{1,1} & x_{1,2} & \dots & x_{1,P} \\ \vdots & \vdots & \vdots & \vdots & \vdots \\ 1 & x_{n,1} & x_{n,2} & \cdots & x_{n,P} \end{bmatrix} \tag{2}$$

So, (1) can also be written in expanded form of a system of linear equations as follows:

$$\begin{matrix} y_1 & = & \beta_0 + \beta_1 x_{1,1} + \beta_2 x_{1,2} + \dots + \beta_P x_{1,P} + \varepsilon_1 \\ y_2 & = & \beta_0 + \beta_1 x_{2,1} + \beta_2 x_{2,2} + \dots + \beta_P x_{1,P} + \varepsilon_2 \\ \vdots & \vdots & \vdots \\ y_n & = & \beta_0 + \beta_1 x_{n,1} + \beta_2 x_{n,2} + \dots + \beta_P x_{n,P} + \varepsilon_n \end{matrix} \tag{3}$$

From (1), the error or the residual is given by

$$\boldsymbol{\varepsilon} = \mathbf{y} - \mathbf{X}\boldsymbol{\beta} \tag{4}$$

From which the mean squared error is given by

$$MSE = \frac{1}{n}(\mathbf{y} - \mathbf{X}\boldsymbol{\beta})^T(\mathbf{y} - \mathbf{X}\boldsymbol{\beta}) \tag{5}$$

Next, we employ the method of Ordinary Least Squares (OLS) to optimized (5) with respect to $\boldsymbol{\beta}$. The alternative to the OLS estimate is the Feasible Generalized Least Squares Estimate treated in [5]. However, the OLS estimates are unbiased and consistent and make a standard error correction for efficiency. We do this by first taking the derivative of MSE with respect to $\boldsymbol{\beta}$ and then setting the result to zero to solve for the estimated values of the parameters and we denote the vector of the determined estimates as beta hat, $\hat{\boldsymbol{\beta}}$, and is given by

$$\hat{\boldsymbol{\beta}} = (\mathbf{X}^{\mathbf{T}}\mathbf{X})^{-1}\mathbf{X}^{\mathbf{T}}\mathbf{y}, \tag{6}$$

where $\hat{\boldsymbol{\beta}}$ is defined as $\hat{\boldsymbol{\beta}} = \begin{bmatrix} \hat{\beta}_0 \\ \hat{\beta}_1 \\ \vdots \\ \hat{\beta}_P \end{bmatrix}$. So, the predicted values $\hat{\mathbf{y}}$ is given as

$$\hat{\mathbf{y}} = \mathbf{X}\hat{\boldsymbol{\beta}} \tag{7}$$

If we now substitute (6) into (7), then we have the predictive model as

$$\hat{\mathbf{y}} = \mathbf{X}(\mathbf{X}^{\mathbf{T}}\mathbf{X})^{-1}\mathbf{X}^{\mathbf{T}}\mathbf{y} \tag{8}$$

If we let the coefficient of the vector $\mathbf{y}$ be denoted by $\mathbf{H}$ and defined as:

$$\mathbf{H} = \mathbf{X}(\mathbf{X}^{\mathbf{T}}\mathbf{X})^{-1}\mathbf{X}^{\mathbf{T}} \tag{9}$$

Then, Eq. (8) becomes the linear equation

$$\hat{\mathbf{y}} = \mathbf{H}\mathbf{y} \tag{10}$$

where **H** is a square matrix that depends on the data values of the features only and not the response variable y and its predictive variable $\hat{\mathbf{y}}$, see [1–3]. **H** is generally referred to as the hat matrix or the influence matrix. It is symmetric and Idempotent. That is, $\mathbf{H}^2 = \mathbf{H}$.

In this presentation, $X_1, X_2, \ldots, X_P$ are the P explanatory variables or features of a cleaned dataset with no particular assumptions about their statistical distribution. However, in the utilization of the method of Ordinary Least Squares in optimizing the $\boldsymbol{\beta}$ values given in (6) assumptions on the conditional expectations of the residuals, variance and covariance with respect to the data values **X** are made. That is, the noise variable $\boldsymbol{\varepsilon}$ has the following properties:

$E[\boldsymbol{\varepsilon}|\mathbf{X}] = \mathbf{0}$; $\mathbf{Var}[\boldsymbol{\varepsilon}|\mathbf{X}] = \sigma^2\mathbf{I}$, that is, that the variance remains constant; and that the covariance of the residuals is zero. That is, $\mathbf{Cov}\ [\boldsymbol{\varepsilon}_i, \boldsymbol{\varepsilon}_j] = 0$ if $i \neq j$. We have further made the following assumptions that the noise or residuals has a joint Gaussian multivariate distribution MVN $(0, \sigma^2\mathbf{I})$ independently of **X** and that the response variable **y** is $\mathbf{y}|\mathbf{X} \sim$ MVN $(\mathbf{X}\boldsymbol{\beta}, \sigma^2\mathbf{I})$. We assume further that $\beta_0 = E\left[\mathbf{y}|\mathbf{X} = \mathbf{0}\right]$ and that each X_i; $i = 1, 2, \ldots, P$, makes a separate contribution to the expected response and they add up without interactions. That is, the contributions the features or explanatory variables makes are linear. Consequently, the rate of change of E[y] with respect of X_i, keeping X_j constant when $i \neq j$, is defined as $\beta_i = \frac{\partial E[y]}{\partial X_i}$ regardless of the point where X_i originates and without mixed terms XiXj. Thus, from these assumptions, we see immediately that $E\left[\hat{\boldsymbol{\beta}}|\mathbf{X}\right] = \boldsymbol{\beta}$. Hence, the Least Squares Estimates of the generalized linear model of (1)'s coefficients are conditionally unbiased regardless of the number of features (explanatory variables), P. The conditional variance of the estimate of parameters is given as:

$$Var\left[\hat{\boldsymbol{\beta}}|\mathbf{X}\right] = \frac{\sigma^2}{n}\left(n^{-1}X^T\mathbf{X}\right)^{-1} \tag{11}$$

From the fitted line given in (10), conditions exist for which this line is either unattainable or results in inaccurate predictions. Conditions such as the existence of collinearity among features or explanatory variables in a dataset. Collinearity involves two or more explanatory variables or features having parallel information. We will deduce here that this definition is similar to the mathematical concepts of linear dependency or linear independency. In model (1) we considered P features but let us instead consider a dataset consisting of the features $\{X_1, X_2, \ldots, X_n\}$, where n is a positive integer. Then, suppose that we are able to find real numbers $\alpha_1, \alpha_2, \ldots, \alpha_n$ such that the following equation holds:

$$\alpha_1 X_1 + \alpha_2 X_2 + \ldots + \alpha_n X_n = 0 \tag{12}$$

If Eq. (12) is true only when all the α_i's; $i = 1, 2, \ldots, n$, are all zero, then $X_1, X_2, \ldots, X_n$ are said to be linearly independent. That is, we cannot express any of the X_i's in terms of one another. That is, $X_i \neq \beta X_j; i \neq j$; β is a real number. On the other hand, if Eq. (12) is true but not all the α_i's are zero, then $X_1, X_2, \ldots, X_n$ are said to be linearly dependent or collinear.

So, why is collinearity a problem? We will explain this by considering the linear model in (1) where β_0 is zero. That is the hat matrix in the fitted model (10) is now of the form $\mathbf{H} = \mathbf{X}(\mathbf{X}^T\mathbf{X})^{-1}\mathbf{X}^T$, where

$$X = \begin{bmatrix} x_{1,1} & x_{1,2} & \cdots & x_{1,p} \\ x_{2,1} & x_{2,2} & \cdots & x_{2,P} \\ \vdots & \vdots & \vdots & \vdots \\ x_{n,1} & x_{n,2} & \cdots & x_{n,P} \end{bmatrix}$$

From this and (11), we can immediately see that the problems lies with inverse matrix $(\mathbf{X}^T\mathbf{X})^{-1}$. We note from linear algebra that by definition:

$$(\mathbf{X}^T\mathbf{X})^{-1} = \frac{Adj(\mathbf{X}^T\mathbf{X})}{\det(\mathbf{X}^T\mathbf{X})} \tag{13}$$

From Linear Algebra, we also know that the inverse in (13) or (11) does not exist when the determinant det $((\boldsymbol{X}^T\boldsymbol{X}))$ of the matrix $(\boldsymbol{X}^T\boldsymbol{X})$ is zero. That is, when the rows or columns of $(\mathbf{X}^T\mathbf{X})$ are collinear. When this happens, the data projector $\mathbf{H}$ – hat matrix is undefined. This is why collinearity is a problem for linear regression models given in (1). Collinearity in a dataset can be identified by visual graphics of scatter plots or by examining the covariance of the features. In general, if

i. n-the number of observations is less than (P + 1), The number of features plus one, we have the case of rank deficiency and hence collinearity exists.

ii. If two of the explanatory or feature variables are proportional to each other, the data is collinear.

iii. If one of the explanatory variables is constant, the dataset is collinear, and

iv. If two of the explanatory variables (features) are otherwise linearly related, the dataset is collinear. In this case, the response variable $y_i = 0$ for each i. That is, the model can be rewritten as

$$\beta_0 + \sum_{j=1}^{P} \beta_j X_j = 0.$$

Conditions (i) – (iv) all amount to saying that det $[\mathbf{X}^T\mathbf{X}] = 0$.

We further explained collinearity and multicollinearity in matrix perspective and mitigation methods to improve machine learning and data analytics algorithms and techniques in terms of eigenvalues and eigenvectors. There by demonstrating the importance of eigenvalues and eigenvectors in modern engineering.

3. Matrix perspective on multicollinearity

Multicollinearity or serial collinearity as used in time series analysis [1–3] implies by definition, as given in Section 2 above, that given $X_1, X_2, \ldots, X_P$ features, where P is a positive integer, there exists constants $a_0, a_1, a_2, \ldots, a_P$, not all zero such that

$$\mathrm{a}_1\mathrm{X}_1 + \mathrm{a}_2\mathrm{X}_2 + \ldots + \mathrm{a}_\mathrm{p}\mathrm{X}_\mathrm{p} = \sum_{i=1}^{P} a_i X_i = \mathrm{a}_0 \tag{14}$$

In the matrix form, if we let

$$\mathbf{a} = \begin{bmatrix} a_1 \\ \vdots \\ a_P \end{bmatrix} \text{ and } \mathbf{X} = (X_1, X_2, \ldots, X_P)^T = \begin{bmatrix} X_1 \\ \vdots \\ X_P \end{bmatrix}, \tag{15}$$

then, multicollinearity implies that

$$\mathbf{a}^\mathrm{T}\mathbf{X} = \mathrm{a}_0; \mathbf{a} \neq \begin{bmatrix} 0 \\ \vdots \\ 0 \end{bmatrix} \tag{16}$$

Taking the variance of both sides of (16), we see that

$$\mathbf{Var}\left[\mathbf{a}^\mathrm{T}\mathbf{X}\right] = \mathbf{Var}[\mathrm{a}_0] = 0. \tag{17}$$

Conversely, if $\mathbf{Var}[\mathbf{a}^\mathrm{T}\mathbf{X}] = 0$, then $\mathbf{a}^\mathrm{T}\mathbf{X}$ must be equal to some constant, say, a_0. Hence, multicollinearity is equivalent to the existence of a nonzero vector **a** such that

$$\mathbf{Var}\left[\mathbf{a}^\mathrm{T}\mathbf{X}\right] = 0 \tag{18}$$

From (18), we note that.
$\mathrm{Var}\left[a^T X\right] = \mathrm{Var}\left[\sum_{i=1}^{P} a_i X_i\right] = \sum_{i=1}^{P}\sum_{j=1}^{P} a_i a_j Cov\left[X_i, X_j\right] = a^T \mathrm{Var}[X] a$. We, therefore have that

$$\mathbf{Var}\left[\mathbf{a}^\mathrm{T}\mathbf{X}\right] = \mathbf{a}^\mathrm{T}\mathbf{Var}[\mathbf{X}]\mathbf{a} \tag{19}$$

We, therefore, say that multicollinearity implies that the equation

$$\mathbf{a}^\mathrm{T}\mathbf{Var}[\mathbf{X}]\mathbf{a} = 0 \tag{20}$$

has a solution $\mathbf{a} \neq \begin{bmatrix} 0 \\ \vdots \\ 0 \end{bmatrix}$.

The Linear Algebra connection is as follows:

We note that $\mathbf{Var}[\mathbf{X}]$ is a P x P square matrix. It is symmetric and positive definite which implies that

$$a^T \mathbf{Var}[\mathbf{X}]\mathbf{a} \geq 0$$

By definition, we know that a matrix B is said to be positive definite, if there exists a matrix C such that

$$C^T B C \geq 0 \text{ or } C B C^T \geq 0$$

Therefore, we have from (14) that

$$\mathbf{a}^{\mathrm{T}}\mathbf{Var}[\mathbf{X}]\mathbf{a} = Var\left[\sum_{i=1}^{P} a_i X_i\right] \geq 0.$$

Since, Var[X] is a P x P symmetric positive definite square matrix, there exists P, (P + 1) (with the constant coefficient included in the model) vectors $V_1, V_2, \ldots, V_P$, the Eigenvectors of Var[X] such that

$$\mathrm{Var}[\mathrm{X}]\mathrm{V_i} = \lambda_i \mathrm{V_i}, \tag{21}$$

where λ_i are the Eigenvalues of the matrix Var[X] corresponding to the Eigenvectors V_i, given $i = 1, 2, \ldots, P$. By the symmetric nature of Var[X], the eigenvalues are non-negative and ordered. That is, $\lambda_1 \leq \lambda_2 \leq \ldots \leq \lambda_P$ with multiplicity $q \leq P$ permitted. That is, the characteristics polynomial $f(\lambda) = b_P\lambda^P + b_{(P-1)}\lambda^{P-1} + \ldots + b_2\lambda^2 + b_1\lambda + b_0$, where $b_1, b_2, \ldots, b_P$ are constants, can be factored as:

$$b_P\lambda^P + b_{(P-1)}\lambda^{P-1} + \ldots + b_2\lambda^2 + b_1\lambda + b_0 = (\lambda-\lambda_1)^q\left(\lambda-\lambda_{(q+1)}\right)\ldots(\lambda-\lambda_P). \tag{22}$$

Thus, given a clean dataset of features **X**, we want to find the eigenvalues and corresponding eigenvectors of the matrix Var[**X**] of the variance of **X**. In order to do this, we shall review some elementary concepts of Linear Algebra. In Linear Algebra, we solve the system of linear eqs.

$\mathbf{AV} = \lambda\mathbf{V}$ by reducing it to a homogeneous system of linear equations:

$\mathbf{AV} - \lambda\mathbf{V} = 0$ or equivalently

$$(\mathbf{A}-\lambda\mathbf{I})\,\mathbf{V} = 0, \tag{23}$$

where **I**, is the identity matrix of the same size with **A**. Clearly, $\mathbf{V} = \begin{bmatrix} 0 \\ \vdots \\ 0 \end{bmatrix}$, if the matrix $\mathbf{A} - \lambda\mathbf{I}$ is not identically zero. So, for any non-zero solutions, linear algebra says that

$$\det\,(\mathbf{A}-\lambda\mathbf{I}) = 0. \tag{24}$$

Where $\det\,(\mathbf{A} - \lambda\mathbf{I})$ is the determinant of the matrix $\mathbf{A} - \lambda\mathbf{I}$. Eq. (24) is a constant coefficient polynomial equation of degree P in λ. That is,

$$b_P\lambda^P + b_{(P-1)}\lambda^{P-1} + \ldots + b_2\lambda^2 + b_1\lambda + b_0 = 0. \tag{25}$$

Let us review some elementary examples of computing eigenvalues and eigenfunctions of square definite matrices. If $\mathbf{A} = \begin{bmatrix} 1 & 0 \\ 0 & 2 \end{bmatrix}$, for example, then $\mathbf{A} - \lambda\mathbf{I} = \begin{bmatrix} 1-\lambda & 0 \\ 0 & 2-\lambda \end{bmatrix}$ and so.

$\det\,(A-\lambda I) = \det\left(\begin{bmatrix} 1-\lambda & 0 \\ 0 & 2-\lambda \end{bmatrix}\right) = \lambda^2 - 3\lambda + 2$, a polynomial of degree 2. The eigenvalues are determined by solving the equations $\lambda^2 - 3\lambda + 2 = 0$. Which gives the values of λ as λ = 1, 2. They are all positive and ordered in the sense that 1 < 2. That is, $1 = \lambda_1 < \lambda_2 = 2$. In the second example, we see the case of repeated or multiple eigenvalues if we let the matrix **A** this time to be $A = \begin{bmatrix} 1 & 2 \\ 0 & 1 \end{bmatrix}$. In this case, we see that

$A–\lambda I = \begin{bmatrix} 1-\lambda & 2 \\ 0 & 1-\lambda \end{bmatrix}$ and $\det(A–\lambda I) = (1–\lambda)^2$. Therefore, we see that if **A** is a P x P matrix and has q repeated eigenvalues, then P – q are non-repeated factors and

$$\det(\mathbf{A}–\lambda\mathbf{I}) = b_q\,(\lambda–\lambda_q)^q\,(\lambda–\lambda_{(q+1)})\ldots(\lambda–\lambda_p).$$

Next, we demonstrate some simple examples for computing eigenvectors corresponding to eigenvalues of a square symmetric positive definite matrix. We will first consider the case of $\mathbf{A} = \begin{bmatrix} 1 & 0 \\ 0 & 2 \end{bmatrix}$, where the eigenvalues were computed to be $\lambda = 1, 2$. Therefore, we need to solve Eq. (23).

$(\mathbf{A} – \lambda\mathbf{I})\ \mathbf{V} = 0$ by letting $\mathbf{V} = \begin{bmatrix} v_1 \\ v_2 \end{bmatrix}$, where v_1 and v_2 are the components of the column vector **V**. So, for $\lambda = 1$, with the given matrix **A**, Eq. (23) becomes on substitution: $\left(\begin{bmatrix} 1 & 0 \\ 0 & 2 \end{bmatrix} - 1\begin{bmatrix} 1 & 0 \\ 0 & 1 \end{bmatrix}\right)\begin{bmatrix} v_1 \\ v_2 \end{bmatrix} = \begin{bmatrix} 0 \\ 0 \end{bmatrix}$. The general solution of which is $\mathbf{V} = v_1\begin{bmatrix} 1 \\ 0 \end{bmatrix}; v_1 \neq 0$. So $\begin{bmatrix} 1 \\ 0 \end{bmatrix}$ is the eigenvector corresponding to the eigenvalue $\lambda = 1$. So, $\lambda = 1$ is associated with the feature $\begin{bmatrix} 1 \\ 0 \end{bmatrix}$, a principal component. For the eigenvalue $\lambda = 2$, we have $A–2I = \left(\begin{bmatrix} 1 & 0 \\ 0 & 2 \end{bmatrix} - 2\begin{bmatrix} 1 & 0 \\ 0 & 1 \end{bmatrix}\right) = \begin{bmatrix} -1 & 0 \\ 0 & 0 \end{bmatrix}$. Therefore, from (23), we solve the system of homogeneous equations $\begin{bmatrix} -1 & 0 \\ 0 & 0 \end{bmatrix}\begin{bmatrix} v_1 \\ v_2 \end{bmatrix} = \begin{bmatrix} 0 \\ 0 \end{bmatrix}$ to get the general solution as $\mathbf{V} = v_2\begin{bmatrix} 0 \\ 1 \end{bmatrix}; v_2 \neq 0$ and the eigenvector corresponding to the eigenvalue $\lambda = 2$ as $\mathbf{V} = \begin{bmatrix} 0 \\ 1 \end{bmatrix}$. So, $\lambda = 2$ is associated with the feature $\begin{bmatrix} 0 \\ 1 \end{bmatrix}$, a second principal component. The dataset information held by each feature is proportional to the totality of the eigenvalues. Thus, $\frac{1}{3} = \frac{1}{1+2}$ of the information is retained by the eigenvector $\begin{bmatrix} 1 \\ 0 \end{bmatrix}$, and $\frac{2}{3} = \frac{2}{1+2}$ is held by the eigenvector $\begin{bmatrix} 0 \\ 1 \end{bmatrix}$. This is the principle behind the method of Principal Component Analysis (PCA) for high dimensional data reduction or feature selection. We will present this method in the section on the Principal Component Analysis. In the PCA method, the eigenvectors are used to scale the data axes. We note that, the eigenvectors are orthogonal to each other. That is, their dot product is zero. From the calculations above, we saw that $\mathbf{V_1} = \begin{bmatrix} 1 \\ 0 \end{bmatrix}$ and $\mathbf{V_2} = \begin{bmatrix} 0 \\ 1 \end{bmatrix}$ and a straight forward computation gives that $\mathbf{V_1} \cdot \mathbf{V_2} = \begin{bmatrix} 0 \\ 0 \end{bmatrix}$.

The eigenvectors $\mathbf{V_i}$'s are selected such that they are normalized (i.e., $\|V_i\| = 1$) and are orthogonal to each other.

That is that

$$\mathbf{V}_j{}^T\mathbf{V}_i = \delta_{ij} = \begin{cases} 1; i{=}j \\ 0; i{\neq}j. \end{cases}$$

In the above examples, we have that $\mathbf{V_1} = \begin{bmatrix} 1 \\ 0 \end{bmatrix}$ and $\mathbf{V_2} = \begin{bmatrix} 0 \\ 1 \end{bmatrix}$. So, we immediately see that $\mathbf{V_1}^T \cdot \mathbf{V_2} = [1\ \ 0]\begin{bmatrix} 0 \\ 1 \end{bmatrix} = 0$, since $1 \neq 2$, and $\mathbf{V_1}^T \cdot \mathbf{V_1} = [1\ \ 0]\begin{bmatrix} 1 \\ 0 \end{bmatrix} = 1$, since $1 = 1$. Clearly, $\mathbf{V_1}^T \cdot \mathbf{V_1} = \|V_1\|^2 = 1$. Which implies that $\|V_1\| = 1$. These are normalized, since we only took the principal components ($v_1 = 1$, and $v_2 = 0$ in $\begin{bmatrix} v_1 \\ v_2 \end{bmatrix}$).

Next, since the eigenvectors span the vector space, any vector can be written as the sum of eigenvectors. That is, for any vector **a**, we have that

$$\mathbf{a} = \sum_{i=1}^{P} \left(\mathbf{a}^{\mathrm{T}}\mathbf{V}_i\right)\mathbf{V}_i \tag{26}$$

So, Var[**X**] can be expressed as Var[**X**] = $\mathbf{VDV}^{\mathrm{T}}$ or $\mathbf{D} = \mathbf{V}^{\mathrm{T}}$Var[**X**]**V**, where **V** is the matrix whose ith column is eigenvector $\mathbf{V}_i$ of the matrix Var[**X**]. So, $\mathbf{V}^{\mathrm{T}}$ is the matrix where $\mathbf{V}_i$ is the ith row, and **D** is a diagonal matrix whose entries are the eigenvalues $\lambda_1, \lambda_2, \dots, \lambda_P$.

Let us suppose that one or more of the eigenvalues are zero. This implies that if $\lambda_1, \lambda_2, \dots, \lambda_q$; $q < P$ are positive eigenvalues, then $\lambda_{(q+1)}, \lambda_{(q+2)}, \dots, \lambda_P$ can be all zero. It follows then that the corresponding eigenvectors $V_{(q+1)}, V_{(q+2)}, \dots, V_P$, all give a linear combination of the features X_i. In this context, a sufficient condition for the existence of multicollinearity in a dataset of features is that the Var[**X**] have zero eigenvalues. Since, det (Var[**X**]) = $\lambda_1 . \lambda_2 . \lambda_3 \dots \lambda_P$.

From the above examples, we saw that the matrix $\mathbf{A} = \begin{bmatrix} 1 & 0 \\ 0 & 2 \end{bmatrix}$ has eigenvalues $\lambda_1 = 1$ and $\lambda_2 = 2$ and its determinant det(**A**) = 2, so it is easy to see that det (**A**) = $\lambda_1 . \lambda_2 = 1.2 = 2$. This is also referred to as the spectral condition or decomposition (Note, that in some literature, eigenvalues are also referred to as spectral values).

On the converse, we suppose that $\mathbf{a}^{\mathrm{T}}$Var[**X**]**a** = 0, and **a** is not a zero vector.

The matrix X of explanatory variables or features of a dataset are multicollinear if and only if the Var[**X**] has zero eigenvalues. Every multicollinear combination of the explanatory variables or features is either an eigenvector of Var[**X**] with zero eigenvalue or a linear combination of such eigenvectors.

These principles are applied in the high dimension data reduction method called the Principal Component Analysis (PCA) and in the Linear Discriminant Analysis for data classification methods. We summarize this section by saying that the eigenvectors of Var[**X**] retains the information contained in **X** proportional to the size of the eigenvalues. That is, the proportion of the eigenvalue λ_i of all the eigenvalues of the matrix **A** is given as: $\frac{\lambda_i}{\sum_{j=1}^{P}\lambda_j}$, see [2, 3]. So, the larger the eigenvalue, the more information is retained by its corresponding eigenvector. Also, collinearity or multicollinearity exists when an eigenvalue of Var[**X**] is zero or very near to zero.

4. Principal component analysis

In this section, we use the methods of eigenvalues and eigenvectors of Section 3 to illustrate the methods of mitigating multicollinearity in high dimension datasets. The inherent difficulties in multivariate linear models for predictive data analysis, regression, and machine learning algorithms are that of visualizing data that has many features or variables and the existence of multicollinearity. Because of these challenges, multivariate data analytics and machine learning methods often begin with some type of dimension simplification and reduction to approximate data in lower dimensional space. Principal component analysis by definition is a rigorous computational method for gaining deeper insights into data and transforming data into lower dimensions to highlight similarities and differences in patterns that may exist. Patterns in data can be hard to identify if the data is of high dimension and visualization tools are not readily available, then the PCA method becomes a very useful tool for analyzing such datasets. In this section, we will go through the steps of performing PCA of cleaned dataset and provide justification of how it works and the role of eigenvalues and eigenvectors.

As we will demonstrate in the section, PCA is a technique that uses the underlying linear algebra principles of Section 3 above to transform a number of correlated or uncorrelated features or variables into smaller number of features called principal components which are ordered according to the size of the eigenvalues of the covariance matrix of the normalized matrix of objects and features of the dataset. This was the original ideas in the PCA method discovered independently by Pearson [3] and Hotelling [6] as a method for describing the variation of a set of multidimensional data in terms of a set of uncorrelated variables in which, each is a linear combination of the original variables [7, 8]. We will show how the variance Var[X] given in (26) of Section 3 and the covariance of the normalized matrix of the objects and features of a dataset are connected with the Singular Value Decomposition (SVD) of data used in PCA applications. Note that normalization of the data is not necessary but must be done if the SVD implementation is to be applied.

This then takes us into the method for implementing PCA with an actual dataset. This demonstration will enable us to see the close connection between PCA and SVD theory of linear algebra. The goal of this section and of the chapter is to explain the eigenvalues and eigenvector theory and practice in the PCA method, and in the following Section 5, provide an example of the use of PCA in mitigating multicollinearity and data dimension reduction in developing a Network Intrusion Detection algorithm for machine learning with the CICIDS2017 network intrusion dataset acquired from the Canadian Institute of Cybersecurity [4].

PCA is applied to a P-dimensional dataset X using the following steps:

i. Determine the normalized P-dimensional dataset's Covariance matrix. Standardizing the feature dataset is not required as proven in the derivation below, but it is useful because most changes of scale are linear transformations of the data that will share the same set of standardized data values [3, 6, 9].

ii. Determine the Covariance matrix's eigenvalues and eigenvectors

iii. Sort the eigenvalues in decreasing order

iv. Select the k eigenvectors corresponding to the k largest eigenvalues, where k is the new feature's subspace dimension. Note that the closer eigenvalue is to zero, the possibility that collinearity exists. By eliminating the eigenvectors corresponding to the eigenvalues closer to zero, leaves the features that have higher contribution to the dataset's best performance.

v. Next, construct the projection matrix from the k eigenvectors you have selected.

vi. Finally, we create a new k-dimensional X feature space by transforming the original **X** dataset.

The pseudocode for computing the PCA with only the covariance of the data set is as follows:

Clean the Dataset X
Compute the dot product matrix: $X^TX = \sum_{i=1}^{n}(X_i - \mu)(X_i - \mu)^T$
Eigenvalues of: $X^TX = VDV^T$
Compute Eigenvectors: $U = XVD^T$
Save the selected number of principal components: $U_P = \{u_1, u_2, \dots, u_P\}$
Compute P features: $Y = XU_P$

Since the CICIDS2017 network intrusion dataset we will be used in Section 5 to implement the PCA method has 79 features and 692,702 objects, so our dataset **X** of Section 3 is now 692,702 x 79 matrix of objects and features. So, we shall let **X** = (X_1, X_2, X_3, ..., X_n), where n = 79 and X_i is as in Section 2, $X_i = (x_{1,i}, x_{2,i}, \dots, x_{n,i})^T, i = 1, 2, \dots, 79$, be the column vectors of **X**. We, therefore, want to transform this matrix, **X** into another matrix, say **Y** of the same dimension 692,702 x 79 as **X**. So, we seek some projector matrix, say **P**, of dimension 79 x 79 such that,

$$\boldsymbol{Y} = \boldsymbol{XP} \tag{27}$$

In the theory of linear algebra, Eq. (27) is a change of basis equation. So, if we let P_1, P_2, ..., P_{79} be the row vectors of **P**, then we have

$$\mathbf{XP} = (X_1P, X_2P, \dots, X_m P) = \begin{bmatrix} X_1P_1 & X_2P_1 & \dots & X_mP_1 \\ X_1P_2 & X_2P_2 & \dots & X_mP_2 \\ \vdots & \vdots & \vdots & \vdots \\ X_1P_{79} & X_2P_{79} & \dots & X_mP_{79} \end{bmatrix} = \mathbf{Y} \tag{28}$$

where $P_i.X_j$ is the standard Euclidean inner (dot) product and m = 692,702. Thus, the original data **X** is projected onto the row vector of **P**. Hence, the columns of **P**, (P_1, P_2, ..., P_{79}) are the new basis for representing the rows of **X**. The columns of **P**, will become the Principal Component directions. We seek the principal components to be independent and in Section 3, we defined independence in terms of the variance of the original data **X**. Thus, the Principal Component tries to uncorrelated the original data **X** by seeking the directions in which the variance is maximized and use them to define the new basis. The statistical definition of a random variable, say, X with a mean, say η, is defined as

$$\sigma^2{}_X = E\left[(X - \eta)^2\right] \tag{29}$$

If we let the centered data be denoted by a vector, say, **r** = $X_i - \eta$ for each *i* = 1, 2, 3, ..., n. Then, the mean of the centered data **r** = (r_1, r_2, ..., r_n) is zero and its variance is given by

$$\sigma_r{}^2 = \frac{rr^T}{n} \tag{30}$$

If **t** = (t_1, t_2, ..., t_n) is another set of measurements with zero mean, then we can generalize the variance idea in (29) and (30) to get the covariance between **r** and **t**. Covariance as we defined in Section 3 and statistically, is a measure of how much two features or variables change together. Thus, variance is a special case of covariance when r is identical with t. If the data is a sample of a sample, we divide in (30) by (n – 1) rather than n. Thus, statistically, we define the covariance of **r** and **t** as:

$$\sigma_{rt}{}^{2} = \frac{rt^{T}}{n-1} \tag{31}$$

If we now let **X** be given as in Section 3 as:

$$X = \begin{bmatrix} x_{1,1} & x_{1,2} & \cdots & x_{1,p} \\ x_{2,1} & x_{2,2} & \cdots & x_{2,P} \\ \vdots & \vdots & \vdots & \vdots \\ x_{n,1} & x_{n,2} & \cdots & x_{n,P} \end{bmatrix} \tag{32}$$

where X_i, $i = 1, 2, \ldots, P$ are the column vectors of (32), then each of the vectors contain all the samples for a particular variable. In this case, we clearly see that X_i is a column vector of the P samples for the ith feature or variable. Thus, the covariance of **X** denoted by Cov[**X**] is the matrix product

$$\text{Cov}[\mathbf{X}] = \frac{\mathbf{XX}^{T}}{n-1} = \frac{1}{n-1}\begin{bmatrix} X_1X_1{}^{T} & X_1X_2{}^{T} & \ldots & X_1X_P{}^{T} \\ X_2X_1{}^{T} & X_2X_2{}^{T} & \ldots & X_2X_P{}^{T} \\ \vdots & \vdots & \ldots & \vdots \\ X_PX_1{}^{T} & X_PX_2{}^{T} & \ldots & X_PX_P{}^{T} \end{bmatrix} \in \mathbb{R}^{PxP} \tag{33}$$

A closed look at the entries in (33) shows that we have computed all the possible covariance pairings between the P-features of variables with the main diagonal entries as the variance of the features and the off-diagonal entries as the covariances. This matrix is called the Covariance Matrix.

Next, we compute the covariance of the matrix **Y** in the reduced space. That is, from (27), we compute

$$\text{Cov}[\mathbf{Y}] = \frac{Y^{T}Y}{n-1} = \frac{(XP)^{T}(XP)}{n-1} = \frac{P^{T}X^{T}XP}{n-1} \tag{34}$$

Let us now let $\mathbf{S} = \mathbf{X}^{T}\mathbf{X}$, then we see that **S** is P x P square symmetric matrix, since from sections 2 and 3, we saw that $(\mathbf{X}^{T}\mathbf{X})^{T} = \mathbf{X}^{T}\mathbf{X}$. With **S**, (34) can be rewritten as:

$$\text{Cov}[\mathbf{Y}] = \frac{P^{T}SP}{n-1}. \tag{35}$$

But, we know from linear algebra that, every square symmetric matrix is diagonalizable with orthonormal or orthogonal matrix. That is, we can find a P x P orthonormal matrix **V** such that $\mathbf{V}^{T}\mathbf{V} = \mathbf{I}$, where **I** is the identity matrix of size **V**, and a P x P diagonal matrix **D** and

$$\mathbf{S} = \mathbf{VDV}^{T} \tag{36}$$

We know from linear algebra that, the columns of **V** are the orthonormal eigenvectors of the matrix **S**, and **D** is a diagonal matrix whose diagonal entries are the eigenvalues of the matrix **S**.

One of the conditions, we saw in Section 3, for the existence of collinearity or multi-collinearity is rank deficiency. That is, if the number of observations are less than the number of features. Under ideal situations, the rank of **S** is the number of its orthonormal eigenvectors. If **S** turns out rank deficient, then one method of mitigation is to complete the rank by generating arbitrary orthonormal vectors to fill the remaining columns of **S**.

This is why we make the mitigation choice for the transformation or projector matrix, P by letting the columns of P be the eigenvectors of **S**, so that, **P** = **V**. On letting **P** = **V** in (35), the derived expression for the covariance matrix Cov[**Y**], we have:

$$\mathrm{Cov}[\mathbf{Y}] = \frac{\mathbf{V}^T\mathbf{V}\mathbf{D}\mathbf{V}^T\mathbf{V}}{n-1}$$

Since, **V** is an orthonormal matrix, $\mathbf{V}^\mathrm{T}\mathbf{V} = \mathbf{I}$, where **I** is the P x P identity matrix. Therefore, for this choice of the projector matrix P, the covariance of the vectors in the new space of features is given as:

$$\mathrm{Cov}[\mathbf{Y}] = \frac{\mathbf{D}}{n-1}. \tag{37}$$

From the computed variances, which are the eigenvalues, we see that the largest variance corresponds to the first Principal Component (PC), the second largest corresponds to the second Principal Component, and so on in the descending order to the very least of the eigenvalues. The information retained by a principal component is proportional to the size of the eigenvalue. Hence, the standard metric for weighing the amount of the information retained by any given PC is the proportion of total variance that it is responsible for. That is, the ratio of its eigenvalue to the total variance given by the formula:

$$\frac{\lambda_i}{\sum_{j=1}^{P}\lambda_j}; i = 1, 2, \dots, P \tag{38}$$

The derivation given here has validated the six steps given above for computing the PCA of a given numerical dataset and a method for organizing the data. That is, we first normalize the dataset and then obtain the covariance and compute its eigenvalues and eigenvectors. Then, sort the eigenvalues in descending order and then obtain the diagonal matrix D by placing the eigenvalues in the descending order diagonally from top left down to the right bottom. Then, we construct the orthonormal matrix, P by putting the corresponding eigenvectors in the same order as the eigenvalues to form the columns of P. That is, we placed the eigenvector obtained with the first and largest eigenvalue in the first column of P, the eigenvector corresponding to the second largest eigenvalue goes into the second column of P, and so on.

As noted in [2, 7, 8, 10], many users of the PCA procedure do retain enough components so as to explain some specified, large percentage of the total variation of the original variables. The total acceptable ratio in percentage values of the eigenvalues is between 70 and 90% but smaller values are also appropriate as n increases [2, 10]. As it was done in [7, 8], in using the CICIDS2017 dataset to implement the PCA method, we will choose the number of components on the basis of a 100% spread or the Empirical rule of large numbers for a spread of three Z-scores so long as the number of selected principal components is less than the number of objects or total recorded network attacks in the dataset.

5. Application of principal component analysis on CICIDS2017 dataset in MATLAB®

In this section, we aim to provide an example of the application of the Principal Components Analysis (PCA) to reducing the dimension of a high dimension dataset in the engineering of Network intrusion detection systems using the MATLAB® PCA library function *pca* [11]. We will use the CICIDS2017 dataset created by the Canadian Institute for Cybersecurity collected in 2017 [4]. The motivation for this example is to demonstrate the role of eigenvalues and eigenvectors in reducing the feature space of a dataset. The use of PCA to model a Network Intrusion Detection System Using Principal Component Analysis algorithm and Decision Tree Classifier as a supervised machine learning linear model was proposed in [8] and in the doctoral dissertation of Oyeyemi Osho in [7] using the CICIDS2017 dataset using the Google Colab development environment and Ski-Learn. Various other approaches have also been developed [12–21]. In this demonstration, we will not implement the pseudocode given in Section 4 above in MATLAB® but as a compromise, we will be using the library function *pca* which was built on the pseudocode.

We will start with a description of the CICIDS2017 data that is from the Canadian Institute for Cyber security (CIC) and University of New Brunswick, Canada collected over several months in 2017. The data was collected eight times in December of 2017 but we will only be using the one for D8.csv file.

5.1 The CICIDS2017 dataset

The CICIDS2017 datasets methods of collection and the aim of the Canadian Institute for Cybersecurity are explained on their website [4]. We will only describe the content of the Data file D8.csv used in this demonstration. It is a 692,703-by-79 excel spreadsheet containing benign and the most up-to-date common attacks that accurately simulate the real-world attacks assembled via packet capture (PCAP). The data include results from using the CICFlowMeter with labeled flows based on the time stamp, source, and destination IPs, source and destination ports, protocols and attack. The D8.csv file has a total of 79 network flow features and 692,703 different attack modes as objects that was imported in the MATLAB® development environment and further cleaned to retain 53 features and 692,703 observations-attack modes. The dataset is too large to display within the script but the file is available on CICIDS website [4].

5.2 PCA of the CICIDS 2017 data D8.csv in MATLAB®

In this demonstration, we will use the MATLAB function *pca* and its arguments to compute the principal component coefficients, also known as the loadings for the cleaned 692,703-by-51 CICIDS2017 dataset matrix will call D8dataset in the MATLAB code [11]. We will determine the number of components required to explain the variance of a fixed percentage of the D8dataset, such as 70% or 95%, create a scree plot of explained variances of the principal components, create a scatter plot of two principal components, create a biplot of two principal components, find the Hotelling's T-Squared statistic for each of the observation (attack mode) in the reduced D8dataset, and obtain the transformed data. We will implement the MATLAB full library function

$$[\text{coeff}, \text{score}, \text{latent}, \text{tsquared}, \text{explained}] = \text{pca}(\text{D8dataset})$$

Next, we will relate the code for pca to the algorithms and pseudocode of Section 4.

D8dataset is the input CICIDS2017 data imported into MATLAB Workspace and stored in the current folder. The data can be cleaned during or after it is imported into MATLAB workspace and stored in the MATLAB current folder. If it is stored in a remote drive, you will have to change drive at running time. D8dataset in this example is a numeric matrix with 692,703 rows and 51 columns. Each row is an observation (object) or sample and each column is a feature or variable. Then, the first step of the PCA sequential steps is that all columns must be zero-centered. That is, the program must have the statement:

$$\text{D8dataset}(:,\text{j}) = \text{D8dataset}(:,\text{j})\text{–mean}(\text{D8dataset}(:,\text{j}));$$

where 'mean' is another one of the many MATLAB library functions, in this case to calculate the mean of column 'j' of the numeric matrix D8dataset.

The MATLAB library function *pca* code will automatically zero-centered and uses the Singular Value Decomposition (SVD) algorithm for fast computing power. However, any reconstructed output will not match. As recommended in Section 4, we will next standardize or normalize the D8dataset by scaling the variance of columns to 1 and mean zero by converting D8dataset to Z-scores as follows:

$$[\text{coeff, score, latent, tsquared, explained}] = \text{pca}(\text{zscore}(\text{D8dataset}));$$

The expressions in the square bracket on the left side of 'pca' are variables holding the output returned by the MATLAB 'pca' function. We will now explain each of them and the data they are holding, but you are free to change them to any expression of your liking that does not alter their content.

coeff: - Is a matrix code variable that holds the Principal Component Coefficients, also known as 'loads' of the 692,703-by-51 dataset numeric matrix D8dataset computed by the MATLAB function *pca*. It is orthonormal and each column is a right singular vector of D8dataset. It is the matrix **V** in the SVD of X in sections 3 and 4. Each column of the coefficient matrix *coeff* contains the coefficients for one principal component. These columns are sorted in descending order of importance of the principal component variance (eigenvalue) with the first column explaining the most variance. *Coeff* is P-by-k numeric matrix where P = size (D8dataset, 2) and k is the number of eigenvectors corresponding to k- eigenvalues. If you specify the "Named-Value" '*Numcomponents*' argument of the *pca* function, then certain conditions apply depending on the rank-deficiency of the input numeric matrix. We will not be using this specification in this example. Note that *size* is also a MATLAB library function that calculate the dimension of a given numeric matrix.

score: - Is a code variable that holds the Principal Component scores of the numeric dataset D8dataset in the reduced features space. It is a n-by-k numeric matrix, where n = *size* (D8dataset, 1). The rows of *score* are observations- cyber-attacks types or modes, and the columns corresponds to components. Again, if you specify the "Named-Value" '*Numcomponents*' argument of the *pca* function, then certain conditions apply depending on the rank-deficiency of the input numeric matrix. We will not be using this specification in this example. If a row i, for example in D8dataset were to be decomposed over the Principal Component vectors, its coefficient is accessed by calling Score(i,j) as:

$$\text{D8dataset(i, :)} = \text{score(i, 1)} * \text{coeff}(:,1) + \text{score(i, 2)} * \text{coeff}(:,2) + \ldots + \text{score(i, p)} * \text{coeff}(:,\text{p})$$

latent: - Is a numeric column vector code variable to hold the Principal Component variances-eigenvalues. That is, the eigenvalues of the covariance of the Z-Scored numeric data matrix D8dataset. It is a numeric column matrix of the length the size of the number of calculated eigenvalues, **k**. When the number of degrees of freedom were to be smaller than size(D8dataset,2) then the *pca* algorithm requires that we specify in the "Name-Value" argument 'Economy = true (default)'. In this case, k equals the number of degrees of freedom. If not then, k equals size(D8dataset,2).

tsquared: - is numeric vector code variable to hold the Hotelling's T-squared statistic [6, 10]. It is the sum of the standardized scores for each observation-cyber-attack mode or type for the D8dataset data. It is a numeric column vector of length size(D8dataset,2). The Hotelling's T-squared statistic is a measure of the multivariate distance of each cyber-attack type from the center of the D8dataset dataset. In this demonstration, we want the Hotelling's T-squared statistic in the reduced space-the space projected by the eigenvectors. So, we will use the Mahalanobis distance. The MATLAB library function for this is 'mahal'. The call syntax for this as follows:

We first compute T-squared statistic using the MATLAB code:

$$[\text{coeff}, \text{score}, \text{latent}, \text{tsquared}] = \text{pca}\ (\text{zscore}(\text{D8dataset}, '\text{NumComponents}', \text{k}, \ldots);$$

We then, compute the T-squared statistic in the reduced space by calling into the program the MATLAB code:

tsqreduced = mahal(score, score); % here tsqreduced is the column numeric vector holding the Hotelling's T-squared of the reduced feature space and then take the difference:

tdifference = tsquared – tsqreduced;

We will use the MATLAB default confidence level of 95% and the degree of freedom d to be one less the number of rows of the cleaned numeric data D8dataset. That is, $d = i - 1$, where i is the number of rows of in D8dataset without missing information. We will use the Named-Value argument *'row'*,*'complete'* in the library function *pca.*

$$[\text{coeff2}, \text{score2}, \text{latent}, \text{tsquared}, \text{explained}, \text{mu2}] = \text{pca}\left(\text{D8dataset}, \ldots\text{'Rows'}, '\text{complete'}\right);$$

explained: - is a numeric column vector program code variable holding the percentage of total variance-eigenvalue explained by each principal component-eigenvector. It is a numeric column vector of length k, the number of eigenvalues of the zscore(D8dataset). We note here that, If the degrees of freedom were to be smaller than size(zscore(D8dataset),2) and we specified Economy = true (default) in the Named-Value argument of *pca*, then k would have been equal to the number of degrees of freedom. If not then, k equals size(zscore(D8dataset),2).
explained = latent/(sum(latent)) *100. This is used in deciding the number of Principal components to keep.

6. The principal component analysis (PCA) workflow in MATLAB

The flow chart below is a MATLAB PCA workflow schema following the step-by-step procedural algorithm of Section 4 implemented on the CICIDS2017 data denoted as D8dataset in **Figure 1**.

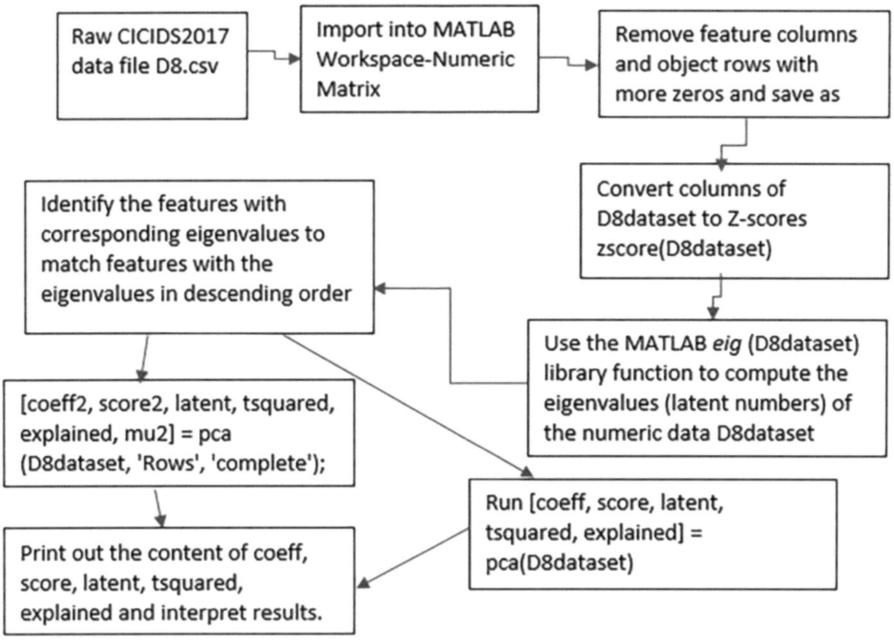

Figure 1.
MATLAB PCA workflow schema.

From the results of the *pca* functions, there are k = 18 non-zero eigenvalues and eigenvectors covering about100% variance. We are going to choose, the first three components for visual analysis and plot score(:, 1), score(:,2), ... , score(:,18) on a 18-dimensional plot to look for clustering along the principal components. If clustering will occur along the *jth* principal component, then we will use the loadings coeff(:,j) to determine which variable explain the clustering.

The MATLAB Script File:

```
% file name – cpad8dataset
% load D8dataset
X = D8dataset(:,1:53);
% Z-score that data
Y = zscore(X);
% calculate the coefficients of the principal components and store into coeff
coeff = cpa(Y);
Disp('coeff = '); disp(coeff);
% This file is too large to display and insert as content of the chapter
% Now we compute both the coefficients, score, and latent-eigenvalues
[coeff, score, latent] = cpa(Y);
% use the MATLAB 'disp' print function to display the results and save for
% insertion in the manuscripts.
% The CICIDS2017 data has a large number of features that are collinear and so it
% cannot be Z-scored. So we must use the D8dataset to find the 53-by-53 numeric
matrix % of coefficients of the principal components coeff.
coeff = pca(D8dataset);
disp('coeff = '); disp(coeff);
```

```
    % Next use the [coeff, score, latent] output format to compute for the score, and
% the latent
    [coeff, score, latent] = pca(D8dataset);
    % As can be seen from the latent column vector, there are 18 non-zero eigenvalues
% and 35 zero eigenvectors. Therefore, pca returned score as 692,702-by-53 numeric
% matrix in the space projected by coeff. The file is too large to insert in manuscripts
    disp('latent = '); disp(latent);
    % The full non-zero latent values in descending order outputted to the variable
% latent is
    %latent 10^15*(8.6625, 1.2393, 0.8758, 0.5336, 0.2340, 0.0425, 0.0412, 0.0133,
% 0.0101, 0.0087, 0.0048, 0.0027, 0.0020, 0.0015, 0.0008, 0.0002, 0.002, 0.002).
    % Next, we calculate the percentage of the explained variance by calling the
% MATLAB function [coeff, score, latent, explained] = cpa(D8dataset) in the
% MATLAB code
    [coeff, score,latent,explained] = pca(D8dataset);
    disp('explained = '); disp(explained);
```

6.1 The T-squared statistic and percent variability explained by principal components

The explained column vector for the non-zero eigenvalues of variances displayed by the MATLAB pca function is:

$$\text{explained} = (80.2355, 11.4724, 4.9404, 2.1664, 0.3934, 0.3817, 0.1232, 0.0931,$$
$$0.0802, 0.0445, 0.0248, 0.0187, 0.0137, 0.0070, 0.0016, 0.0009, 0.0004, 0.0001)$$

This shows that the first 18 principal components of the features given in the appendix explain 100% of all variability. So, the dimension of CICIDS D8dataset data can be reduced from the cleaned 51 features to 18 features. This turns out to be the top 18 on the list in the Appendix **Table A1**. The first three components explained 96.6483% of all variability as shown in Appendix **Table A2** and the two features in Appendix **Table A1** were removed.

We can now visualize the data representation in the space of the first three principal component with the MATLAB 3-D library function scatter3 plot using the reduced score space as:

```
Scatter3 (score(: , 1), score(: , 2), score(: , 3))
        axis equal
        xlabel('PC1')
                ylabel('PC2')
                        zlabel('PC3')
```

The 3-D scatter plot produced is given in **Figure 2** below.

In **Figure 2**, we see clearly that the largest variability is along the first principal component axis (PC1). It is the largest possible variance among all possible choices of the first axis. We also see that the variability along the PC2 axis is the largest among all possible remaining choices of the second axis and the PC3 axis has the third largest variability, which is significantly smaller than the variability along the PC2 axis. We see from explained vector that, the fourth through eighteenth principal component

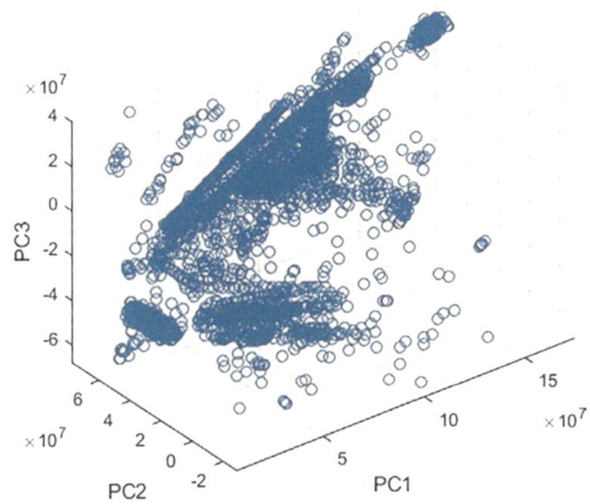

Figure 2.
3-D Scatter Plot of PC1, PC2, and PC3.

axes we do not have to necessary inspect, because they explain only 3.3517% of all variability in the data which is within the margin of the Hotelling's T-squared statistic error [6, 10] of 95% confidence interval.

The Hotelling's T-squared statistic for all the components is calculated by the MATLAB pca function and stored in the column vector code variable tsquared of the modified new line in the MATLAB script file given as:

$$[\text{coeff}, \text{score}, \text{latent}, \text{tsquared}, \text{explained}] = \text{pca}(Z); \%Z \text{ is a } 692,703 - \text{by}$$
$$\% - 51 \text{ cleaned D8dataset to further removed the unwanted features.}$$

Using the MATLAB disp(tsquared) function to print tsquared and affirmed that PC1, PC2, and PC3 are the dominant components.

6.2 Bi_Plot of the first two principal component PC1 and PC2

We will now create and center the data and store it in the vector code variable Zcentered with the MATLAB code

$$\text{Zcentered} = \text{score}^{*}\ \text{coeff}'$$

The original CICIDS 2017 D8dataset data is centered by subtracting the column means from corresponding columns. We now want to visualize both the orthonormal principal component coefficients for each variable and the principal component scores for each observation in a single plot. We do this with the MATLAB library function *biplot*.

$$\text{biplot}\left(\text{coeff}(:,1:2), '\text{scores}', \text{score}(:,1:2), '\text{varlabels}', \{'\text{v\_1}', '\text{v\_2}', '\text{v\_3}', '\text{v\_4}'\}\right);$$

Here we will demonstrate with only the first two principal components, Principal Component 1 and Principal Component 2 and 51 variables in the reduced feature space explained by the principal components as shown in **Figure 3**.

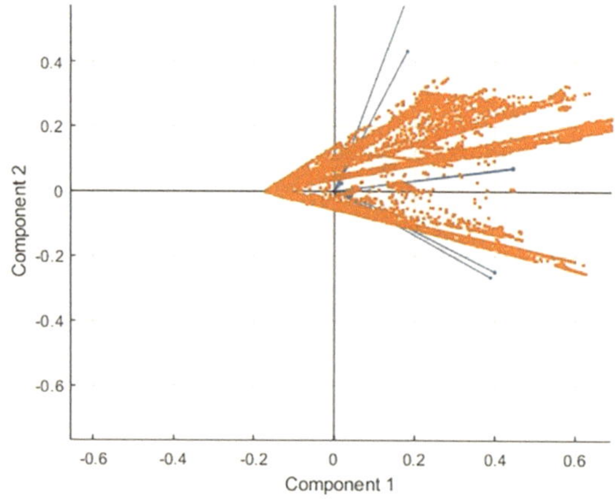

Figure 3.
A 2-D Biplot of PC1 Vs PC2.

All variables are represented in this biplot by a vector, and the direction and length of the vector indicate how each variable contributes to the two principal components in the plot. For example, the first principal component, which is on the horizontal axis, has positive coefficients for the variables in blue lines. Therefore, vectors are directed into the right half of the plot. The largest coefficient in the first principal component is the vector, corresponding to the blue line above all blue lines.

The second principal component, which is on the vertical axis, has negative coefficients for the blue lines to the right below the component 1 axis, and positive coefficient for other variables as shown in **Figure 3**.

This 2-D biplot also includes a point for each of the 692.703 observations, in red, with coordinates indicating the score of each observation for the two principal components in the plot. Pointing near the left edge of the plot have the lowest scores for the first principal component. The points are scaled with respect to the maximum score value and maximum coefficient length, so only their relative locations are determined from the plot in **Figure 3**.

6.3 The impact of PCA on classifier performance: Decision tree classifier in MATLAB for the network intrusion CICIDS2027 dataset

In Section 2, we gave a multivariate linear model of a dataset $\mathbf{X}$ in (10) *as* $\hat{\mathbf{Y}} = \mathbf{HY}$, where $\mathbf{Y}$ is the response variable and $\mathbf{H} = \mathbf{X}(\mathbf{X}^T\mathbf{X})^{-1}\mathbf{X}^T$ is the projector matrix called the hat-matrix. We saw that the sample variable $\mathbf{Y}$, depending on the behavior of the hat matrix $\mathbf{H}$ which depends on whether the covariance matrix of the data $\mathrm{Cov}[\mathbf{X}] = \mathbf{X}^T\mathbf{X}$ has an inverse or not, may not be predicted well by $\hat{\mathbf{Y}}$. This happens when the inverse of Cov[X] does not exists. If it does not have an inverse then there are some features of the dataset $\mathbf{X}$ that might be collinear. In Section 3, we developed the theory of eigenvalues and eigenvectors and showed that collinearity or multicollinearity exist if the determinant of the Cov[$\mathbf{X}$] is zero and this happens when the covariance matrix Cov[$\mathbf{X}$] has a zero eigenvalue or near zero eigenvalues. In Section 4, this information is used to develop the Principal Component Analysis (PCA) algorithm for mitigating collinearity or multicollinearity and retained the explained principal components. In this section, we implemented the PCA using its MATLAB library

Training results	Validation accuracy	Validation total cost	Validation error rate	Prediction speed	Training time	Model size
Fine decision tree with PCA	97%	18,968	3.0%	710,000 obs/sec	25.9 sec	45 KB
Fine decision tree without PCA	99.7%	2007	0.3%	~200,000 obs/sec	62.652 sec	54 KB

Table 1.
Models training results.

function *pca* for the Canadian Institute of Cybersecurity CICIDS2017 dataset that has 692,703 observations of six different types of network attacks and 79 features on the OptiPlex 5090 DELL Intel Core *i*7 desk top computer in parallel. The PCA retained 18 uncorrelated explained principal components and the dimension of the dataset reduced to 692,703 – by-53. Our next step is to assess the impact PCA has on data classification classifiers with the MATLAB Fine Decision Tree Classifier library function *fitctree* in the train decision trees Classification Learner App to the accuracy and speed in training and classifying the different six different Network attacks in the CICIDS2017 dataset with the PCA dataset and the non-PCA dataset.

We classify the original dataset D8data.csv with the MATLAB Fine Decision Tree Classifier uncleaned with the 692,703 observations and 78 features and six Response Classes: BENIGN, DoS GoldenEye, DoS Hulk, DoS Slowhttptest, DoS slowloris, and Heartbleed using 90% of the data for training and 10% of the data for testing. Then, the same approach with the reduced principal component data consisting of 692,703 observations and the 18 principal components that covers 99.8% explained variance. The dataset was also divided into 90% for training and 10% for testing. The percentage variance explained was set for greater than or equal to 95%. The training results are given in **Table 1** above.

Both models utilized 623,433 of the 692,703 observations and a test data of 69,270 observations. The Fine Decision Tree Classifier without PCA selected all the 78 Individual features to cover the 95% variance while the PCA Fine Decision Tree Classifier selected 4 of the 18 features within the 95% explained Principal Components variance. As can be discerned from **Table 1**, PCA has enhanced the training time, prediction speed, model size compression of the MATLAB Fine Decision Tree Classifier algorithm and traded off in validation accuracy, total cost and error rate.

7. Conclusions

Our aim in this chapter is to bring to light the role eigenvalues and eigenvectors theory are put into practical use in the era of modern engineering that increasingly is dependent on artificial intelligence and machine learning from a data centric perspective. We hope to have presented it in a way that anybody with a limited knowledge of mathematics and statistics, particularly in linear algebra will be able to follow. We showed why multicollinearity is a problem in multivariate linear modeling of complex data and its mitigation with the Principal Component Analysis method where eigenvalues and eigenvectors are prominently featured. A detailed derivation of the PCA method and an example of its application with one of the most complex data structures in Network Intrusion data from the Canadian Institute of Cybersecurity was presented.

Acknowledgements

I wish to thank the anonymous reviewer who was professional and graciously gave constructive feedback and valuable suggestions that have greatly improved this chapter both in content and presentation. I also wish to extend my gratitude to the Publishers for being patient with me during this process.

Appendix

See **Table A1**.

Features	Features	Features	Features	Features
Destination Port	Bwd Packet Length Std	Bwd IAT Mean	Down/Up Ratio	Idle Min
Flow Duration	Flow Bytes/s	Bwd IAT Std	Average Packet Size	
Total Fwd Packets	Flow Packets/s	Bwd IAT Max	Avg Fwd Segment Size	
Total Backward Packets	Flow IAT Mean	Bwd IAT Min	Avg Bwd Segment Size	
Total Length of Fwd Packets	Flow IAT Std	Fwd Header Length	Fwd Header Length	
Total Length of Bwd Packets	Flow IAT Max	Bwd Header Length	Subflow Fwd Packets	
Fwd Packet Length Max	Flow IAT Min	Fwd Packets/s	Subflow Fwd Bytes	
Fwd Packet Length Min	Fwd IAT Total	Bwd Packets/s	Subflow Bwd Packets	
Fwd Packet Length Mean	Fwd IAT Mean	Min Packet Length	Subflow Bwd Bytes	
Fwd Packet Length Std	Fwd IAT Std	Max Packet Length	Init_Win_bytes_forward	
Bwd Packet Length Max	Fwd IAT Max	Packet Length Mean	Init_Win_bytes_backward	
Bwd Packet Length Min	Fwd IAT Min	Packet Length Std	act_data_pkt_fwd	
Bwd Packet Length Mean	Bwd IAT Total	Packet Length Variance	min_seg_size_forward	

Table A1.
List of reduced data features in the CICIDS 2017 D8.csv dataset.

See **Table A2**.

PC	Eigenvalue $1*10^{15}$	% Explained	PC	Eigenvalue	% Explained $1*10^{15}$
Destination Port	8.6625	80.2355	Bwd Packet Length Max	0.0048	0.0248
Flow Duration	1.2393	11.4724	Bwd Packet Length Min	0.0027	0.0187
Total Fwd Packets	0.8758	4.9404	Bwd Packet Length Mean	0.002	0.0137
Total Backward Packets	0.5336	2.1664	Bwd Packet Length Std	0.0015	0.007
Total Length of Fwd Packets	0.234	0.3934	Flow Bytes/s	0.0008	0.002
Total Length of Bwd Packets	0.0425	0.3817	Flow Packets/s	0.0002	0.0016
Fwd Packet Length Max	0.0412	0.1232	Flow IAT Mean	0.0002	0.0009
Fwd Packet Length Min	0.0133	0.0931	Flow IAT Std	0.0002	0.0004
Fwd Packet Length Mean	0.0101	0.0802	Flow IAT Max	0.0000	0.0001
Fwd Packet Length Std	0.0087	0.0445			

Table A2.
100% Explained by 18 Principal Components were the top 18 on the list.

A.1 Removed unwanted features

Flow Bytes/s, Flow Packets/s

Author details

Tor A. Kwembe
Jackson State University, Jackson, Mississippi, USA

*Address all correspondence to: tor.a.kwembe@jsums.edu

References

[1] Carter HR, Griffiths WE, Lim GC. Principles of Econometrics. 5th ed. New York: Wiley; 2018. ISBN: 9781119510567

[2] Everitt BS, Dunn G. Applied Multivariate Data Analysis, John Wiley & Sons ltd. 2nd ed. Chichester, West Sussex, UK: John Wiley; 2001. ISBN: 978-0-4707-1117-0

[3] Pearson K. 1901 on lines and planes of closest fit to systems of points in space. Philosophical Magazine. 1901;**2**:559-572. DOI: 10.1080/14786440109462720

[4] Intrusion detection evaluation dataset, Canadian Institute for Cybersecurity. Available from: https://www.unb.ca/cic/datasets/ids-2017.html

[5] Beck N, Katz JN. What to do (and not to do) with time-series-cross-section data in comparative politics. American Political Science Review. 1995;**89**(3): 634-647

[6] Hotelling H. Analysis of a complex of statistical variables into principal components. Journal of Education & Psychology. 1933;**24**(417–441):498-520. DOI: 10.1037/h0071325

[7] Osho O. Network intrusion detection system using principal component analysis and linear discriminant analysis [PhD dissertation]. Jackson, Mississippi, USA: Jackson State University; 2022.

[8] Osho O, Hong S, Kwembe TA. Network intrusion detection system using principal component analysis algorithm and decision tree classifier. In: Proceedings of the 2021 International Conference on Computational Science and Computational Intelligence (CSCI). 2021. pp. 273-279. DOI: 10.1109/CSCI54926.2021.00117

[9] Jolliffe IT, Cadima J. Principal component analysis: A review and recent developments. Philosophical Transactions of the Royal Society A. 2016;**374**:20150202. DOI: 10.1098/rsta.2015.0202

[10] Jackson JE. 2003 a user's Guide to Principal Components. New York: Wiley; 2003. ISBN: 978-0-471-47134-9

[11] MATLAB-Mathworks. Available from: https://www.mathworks.com

[12] Bouzida Y, Cuppens F, Cuppens-Boulahia N, Gombault S. Efficient Intrusion Detection Using Principal Component Analysis. In 3 ème conference sur la sécurité et Architectures RéseauxSAR. LaLonde, France; June 2004

[13] Glass-Vanderlan TR, Iannacone MD, Vincent MS, Chen Q, Bridges RA. A survey of intrusion detection systems leveraging host data. 2018. arXiv: 1805.06070 [CS. CR]. [Online]. Available from: http://arxiv.org/abs/1805.06070 [Accessed: November 22, 2021]

[14] Mechtri L, Tolba FD, Ghoualmi N. Intrusion detection using principal component analysis. In: 2nd International Conference on Engineering System Management and Applications. 2010. pp. 1-6

[15] Mishra A, Cheng AML, Zhang Y. Intrusion detection using principal component analysis and support vector machines. In: Proceedings of the IEEE 16th International Conference on Control & Automation (ICCA) 9-11 October 2020; Virtual. pp. 907-912. DOI: 10.1109/ICCA51439.2020.9264568

[16] Sharafaldin I, Lashkari AH, Ghorbani AA. Toward generating a new intrusion detection dataset and intrusion

traffic characterization. In: Proceedings of the Fourth International Conference on Information Systems Security and Privacy. Funchal, Madeira, Portugal. 2018. pp. 108-116. DOI: 10.5220/0006639801080116

[17] Sharma A, Paliwal KK. Linear discriminant analysis for the small sample size problem: An overview. International Journal of Machine Learning and Cybernetics. 2015;**6**(3): 443-454. DOI: 10.1007/s13042-013-0226-9

[18] Tharwat A, Gaber T, Ibrahim A, Hassanien AE. Linear discriminant analysis: A detailed tutorial. AL Communications. 2017;**30**(2):169-190

[19] Xanthopoulos P, Pardalos PM, Traflis BT. Robust Data Mining–Linear Discriminant Analysis. New York: Springer; 2013. pp. 27-23. DOI: 10.1007/978-1-4419-9878-1

[20] Zhang B, Liu Z, Jia Y, Ren J, Zhao X. Network intrusion detection method based on PCA and Bayes algorithm. Security and Communication Networks. 2018;**208**:11. DOI: 10.1155/2018/1914980

[21] Zhong R, Liu S, Li H, Zhang J. Robust functional principal component analysis for non-Gaussian longitudinal data. Journal of Multivariate Analysis. 2022;**189**:104864. DOI: 10.1016/j.jmva.2021.104864

Chapter 8

Principal Components and Factor Models for Space-Time Data of Remote Sensing

Carlo Grillenzoni

Abstract

Time-lapse videos, created with sequences of remotely-sensed images, are widely available nowadays; their aim is monitoring land transformations, both as regards natural events (e.g., floods) and human interventions (e.g., urbanizations). The corresponding datasets are represented by multidimensional arrays (at least 3-4D) and their spectral analysis (eigenvalues, eigenvectors, principal components, factor models) poses several issues. In particular, one may wonder what are the statistically meaningful operations and what is the treatment of the space–time autocorrelation (ACR) across pixels. In this article, we develop principal component analysis (PCA, useful for data reduction and description) and factor autoregressive models (FAR, suitable for data analysis and forecasting), for 3D data arrays. An extensive application, to a real case study of a Google Earth video, is carried out to illustrate and check the validity of the numerical solutions.

Keywords: autoregressive models, eigenvalues space-time, least squares, multidimensional arrays, space-time forecasting

1. Introduction

Modern remote sensing technologies, for data acquisition and processing, provide large amounts of environmental data, with good coverage in space and time. When such data are in the form of sequences of digital images, properly georeferenced and equalized, then an entire timelapse video can be constructed. These movies allow dynamic monitoring and surveillance of earth areas subject to natural events (such as floods, landslides, and wildfires) and human interventions (such as urbanization, agriculture, and wars). A classical example is the Google Earth platform which edits videos from the imagery of LandSat and Copernicus satellites and broadcasts them through its YouTube channel. Recently, [1] has also implemented an online engine that enables users to build their videos at a global scale; it is continuously improved as regards space–time resolution and image quality.

Apart from descriptive and entertaining aspects, timelapse videos are useful for monitoring and surveillance purposes, to signal land hazards and risks. In this perspective, numerical methods for representing the video frames and for obtaining meaningful information are crucial. From a statistical viewpoint, the video datasets are multidimensional (4D) arrays of space–time positive numbers; given their complexity, the application of dimensional reduction techniques, such as principal components (PC) and factor analysis, is necessary. Basic algebraic instruments are the eigenvectors of the covariance matrices and their projection properties in the space. These techniques are well known for single digital images; e.g., [2] apply PCA to image compression, [3] to object detection and image segmentation, and [4] to heterogeneous geodata layer fusion.

Point spatial data have a smaller size than images and allow for a formal treatment of the temporal component; e.g., [5] uses PC for directional (ridge) clustering of earthquake epicenters, [6] define a PCA approach in the attribute space that maintains the data structure in the spatiotemporal domain, and [7] develop space–time PCA toward functional statistical analysis. Operationally, [8] model networks of environmental stations as a multivariate AR system and use PC for reducing its dimension; they also study the effect of temporal ACR on PC extraction.

PCA of human videos has been considered in Ref. [9]; given the heterogeneity of scenes, the main goal is clustering the frames in homogeneous groups for subsequent uses (e.g., clip extraction). Liu et al. [10] used a nonlinear version of PCA to reach a more operational goal of automatic video editing. PCA methods for semantic video interpretation, to be applied in computer vision and robotics, have a long history mostly based and supervised classification; see Ref. [11] and reference therein. They require the construction of large and consistent datasets of annotated (human pre-classified) frames and sequences. Similarly, [12] using *neural* classifiers have tried to forecast video sequences out-of-sample, i.e., beyond the observed interval. This attempt is computationally demanding as requires the calibration of complex neural networks, which are over-parameterized models from a statistical viewpoint.

In this paper, we consider timelapse videos of remote sensing data and use principal components both for synthesis and forecasting. In the time domain, PCs may resume video frames as long exposure photography, to have an instantaneous view of the land change. In the space domain, PCs may resume local series and reduce the dimension of space-time systems, to implement simpler factor models. We show that the presence of ACR is an issue from the theoretical viewpoint for PC estimation, but has limited practical effects both on data description and parameter estimates. We compare the forecasting performance, on out-of-sample frames, of factor AR models (that may be modeled as univariate time series) and space-time AR models (that are similar to multivariate systems; see [13].

The paper is organized as follows: Section 2 deals with PCs in the time domain as a general tool of frame synthesis, compared to the simple arithmetic averaging. Section 3 discusses PCs in the spatial domain as a tool for building factor models; here, least squares (LS) estimator and forecasting algorithms are developed. Throughout, an extended numerical application to the Google Earth video of the Iquitos city (Peru) in the period 1984–2022 is carried out to illustrate and compare the methods.

2. Principal component analysis of videos

Remote sensing technologies and digital image processing generate numerical data on regular lattices. Typical datasets are in the form of 4D arrays of the type

$\mathbb{Y} = \{Y_{ijlt}\}$, where $i = 1, 2 \dots n, j = 1, 2 \dots m$ are indices of pixel position (which may be transformed into latitude and longitude), $l = 1, 2 \dots k$ are the spectral bands (k = 3 for RGB color images), and $t = 1, 2 \dots T$ is the index of time (daily, monthly, and annual). The first approach of dimensional reduction is transforming the spectral bands into a single indicator, as the grayscale in the visible range, or normalized difference vegetation index (NDVI) in the near-infrared channel, to obtain 3D arrays. The resulting values Y_{ijt} are usually autocorrelated (in space and time) and non-stationary (with spatial and temporal trends).

The PC analysis of classical (2D) data matrices $\boldsymbol{X} = \{X_{ij}\}$, with N units and M variables, $N > M > 3$, is a technique of dimensional reduction to obtain a few linear combinations of the columns $\boldsymbol{x}_j$ which capture most of the variability and allow visualization in 1-3D space. A direct application to image processing is to compress a color picture $\mathbb{X} = \{\boldsymbol{X}_l\}_1^3$ in its grayscale version, e.g., [2, 14]. The PCA technique vectorizes the RGB layers $\boldsymbol{x}_l = \text{vec}(\boldsymbol{X}_l)$, builds a $nm \times 3$ matrix $\boldsymbol{\mathcal{X}} = [\boldsymbol{x}_1, \boldsymbol{x}_2, \boldsymbol{x}_3]$, estimates the first principal component $\boldsymbol{z}_1 = \sum_{l=1}^{3} v_{l1}\boldsymbol{x}_l$, and reshapes it as a new image $\boldsymbol{Z}_1$. The questions now are: How can PCA be applied to a video $\mathbb{Y} = \{\boldsymbol{Y}_t\}_1^T$, with T frames, and what meaningful results may it produce?

Authors in Ref. [4] applied PCA to a 3D matrix with k = 9 geographic layers; the goal is to improve the GIS overlaying technique based on the arithmetic mean, which uses uniform weights $v_l = 1/k$. The fundamental step, before applying PCA, is rescaling the layers in the range [0,1], with the transformation $X_{ijl}/\max_{ij}(X_{ijl})$; the PCA technique then provides non-uniform weights which enhance the most significant layers. Now, even in remote sensing sequences $\{\boldsymbol{Y}_t\}$ there is the goal of spatial mapping, but with the most ambitious purpose of representing the time-evolution of the phenomenon in a single frame (as in long-exposure photography). In this context, there are no problems of data heterogeneity, as the frames belong to the same band; rather, there is an issue of strong space-time ACR.

The steps for PCA investigation of a video are as follows: transform the initial 4D dataset in grayscale (frame by frame) or proceed separately on each color band. Vectorize the resulting 3D array $\mathbb{Y} = \{Y_{ijt}\}$, $n \times m \times T$, as

$$\boldsymbol{\mathcal{Y}} = [\text{vec}(\boldsymbol{Y}_1), \text{vec}(\boldsymbol{Y}_2) \dots \text{vec}(\boldsymbol{Y}_t) \dots \text{vec}(\boldsymbol{Y}_T)], \tag{1}$$

obtaining a 2D matrix of size $nm \times T$. Consider its *centered* (mean zero) version

$$\boldsymbol{\mathcal{Y}}_0 = (\boldsymbol{\mathcal{Y}} - \mathbf{1}_{nm} \otimes \bar{\boldsymbol{y}}), \qquad \bar{\boldsymbol{y}} = \mathbf{1}'_{nm}\boldsymbol{\mathcal{Y}}/nm, \tag{2}$$

where $\mathbf{1}_{nm}$ is a unit vector of length nm. Compute the covariance matrix $\boldsymbol{C}$ (which is symmetric and positive definite) and perform its spectral factorization

$$\boldsymbol{C} = \boldsymbol{\mathcal{Y}}'_0\boldsymbol{\mathcal{Y}}_0/nm, \qquad \boldsymbol{C} = \boldsymbol{V}\Lambda\boldsymbol{V}', \tag{3}$$

where $\boldsymbol{V}, \Lambda$, are $T \times T$ matrices of eigenvectors and eigenvalues, where the latter are placed in decreasing order: $\lambda_k \geq \lambda_{k+1}$ within Λ.

Now, the meaningful first PCs of the space–time array $\mathbb{Y}$ are given by

$$\boldsymbol{z}_0 = \boldsymbol{\mathcal{Y}}_0\boldsymbol{v}_1, \quad \text{projection on the first PC axis,} \tag{4}$$

$$\boldsymbol{z}_1 = \boldsymbol{\mathcal{Y}}\boldsymbol{v}_1/\|\boldsymbol{v}_1\|_1, \quad \text{weighted average of } \boldsymbol{Y}_t, \tag{5}$$

where $\boldsymbol{v}_1$ is the first eigenvector of the orthogonal matrix $\boldsymbol{V}$, and $\|.\|_1$ is the absolute norm. Finally, the vectors $\boldsymbol{z}_0, \boldsymbol{z}_1$ must be reshaped as $n \times m$ matrices $\boldsymbol{Z}_0, \boldsymbol{Z}_1$ and encoded in uint8 format [0,255], to be represented and processed as images. While $\boldsymbol{Z}_1$ provides a weighted average of the frames, the image $\boldsymbol{Z}_0$ is more essential and may detect the major changes of the sequence $\{\boldsymbol{Y}_t\}$.

An issue in this approach, compared to the classical PCA, is the lack of independence of data. The space–time ACR of pixels may induce bias and inefficiency in the estimates; in particular, in the standard errors of the eigenvector $\boldsymbol{v}_1$, see [15, 16]. As in regression models, a naïve method to improve the statistical properties is to include "lagged" terms into the system; in the above framework, this means computing the matrix $\boldsymbol{V}$ in Eq. (3) on the *augmented* array

$$\mathcal{Y}^* = [\mathcal{Y}, \mathcal{Y}_1], \qquad \mathcal{Y}_1 = \text{vec}(\mathbb{Y}_1), \quad \mathbb{Y}_1 = \{Y_{i-1,j-1,t-1}\}, \tag{6}$$

where the lagged array $\mathbb{Y}_1$ is integrated with missing terms, e.g., putting the column $\boldsymbol{y}_{m,t-1} = \boldsymbol{y}_{1,t-1}$. The resulting matrix (6) has size $nm \times T(T-1)$, and only the first T elements of $\boldsymbol{v}_1^*$ are used for computing the PCA vectors $\boldsymbol{z}_0, \boldsymbol{z}_1$ in Eqs. (4) and (5).

As in time series, e.g., [8], a substantial reduction of ACR is provided by the space–time differencing $y_{ijt} = (Y_{ijt} - Y_{i-1,j-1,t-1})$. Since y_{ijt} also assume negative values, the nature of the implied coefficients $\boldsymbol{v}_1$ substantially changes, and they may not be suitable for the original data Y_{ijt}. Furthermore, reconstructing the target image $\boldsymbol{Z}$ from the PCA image $\boldsymbol{z} = \{z_{ij}\}$ of the series y_{ijt} is difficult and biased. Indeed, this requires the spatial integration $Z_{ij} = Z_{i-1,j-1} + z_{ij}$, which in turn involves $n + m - 1$ initial values Z_{i1}, Z_{1j}; these border values are arbitrary and may distort the entire image $\boldsymbol{Z}$.

Finally, for point (non-lattice) data, with matrices $\boldsymbol{X}_t = \{x_{skt}\}$, $N \times M$, equispaced in time t but irregularly distributed in space with coordinates (i_s, j_s), [17] and [6] have considered a PCA approach based on a spatially weighted covariance matrix, as in the Moran index

$$\boldsymbol{C}_W = \frac{1}{NT}\sum_{t=1}^{T}(\boldsymbol{X}_t - \overline{\boldsymbol{x}})'\boldsymbol{W}_N(\boldsymbol{X}_t - \overline{\boldsymbol{x}}), \qquad w_{ij} = \begin{cases} 0, & i = j, \\ 1, & \text{sparse}, \end{cases}$$

where $\boldsymbol{W}$ is a $N \times N$ contiguity matrix of the points based on the assumption of interactions (e.g., nearest neighbors). The derivation of the matrix $\boldsymbol{W}$ for lattice data is possible using geometrical rules of chess moves (e.g., rook, queen, etc.); in the presence of asymmetry, the positive definiteness of $\boldsymbol{C}$ is preserved by

$$\boldsymbol{C}_W = \frac{1}{2nm}\mathcal{Y}_0'(\boldsymbol{W}_{nm} + \boldsymbol{W}_{nm}')\mathcal{Y}_0.$$

However, apart from the arbitrariness of the contiguity rules, for image data the building and use of the $nm \times nm$ array $\boldsymbol{W}$ is numerically demanding for lattices [18], even in the lowest resolution case (n, m) = (144, 256).

Anyway, the presence of ACR mostly affects the standard errors and test statistics of the estimates $\boldsymbol{v}_1$ (see [15] p. 299); hence, it may be a minor problem when using the PCs for image representation and processing. Instead, the mentioned corrections may introduce significant bias; thus, in the application, we mainly focus on Eqs. (4) and (5) for image synthesis.

2.1 Application to Google earth video

Google [1] creates timelapse videos based of LandSat and Copernicus satellite images, which are properly georeferenced and homogenized. The service is global, locally zoomable, and enables to evaluate how the Earth has changed over the past 40 years (since 1984). A set of high-resolution videos of interesting areas are put on the YouTube platform and can be downloaded; we consider the Iquitos city in Peru, located on the banks of Rio Amazonas, see [19]. As a consequence of the periodic floods, the change of the river bed between 1984 and 2022 is impressive, as well as the impact on urban growth (see **Figure 1**).

Given the computational load of the algorithms for a laptop computer with MATLAB software, we consider a low-resolution video (240p) and an area of about 18 $\times$ 22 Km. This yields a data array in black and white of size n = 288, m = 368, T = 39, which provides N = 4,133,376 observations. The color display is shown in **Figure 1**; notice the significant displacement of the river bed during 39 years.

Figure 2 provides the main statistics of the video in grayscale; the time-trends in Panel 2a show a rough 5-year cycle of the vegetation activity, which may be related to El Niño oscillation (ENSO). Panel 2b provides the correlation matrix of Eq. (3): $\boldsymbol{R} = \boldsymbol{S}^{-1/2}\boldsymbol{C}\boldsymbol{S}^{-1/2}$ with $\boldsymbol{S} = \boldsymbol{C} \odot \boldsymbol{I}$ and shows that nearest frames are more correlated. Panel 2c shows the path of the first eigenvectors of Eqs. (3) and (6) and differences series y; the first two are proportional (and coincide when are normalized by $1/\|\boldsymbol{v}_1\|_1$), whereas the third is consistent with the first. This means that ACR has *not* a great effect on the estimates of $\boldsymbol{v}_1$. Finally, Panel 2d displays the relative eigenvalues $\lambda_k/\text{tr}(\Lambda)$ in the original and differenced series; it shows that PC1 is more significant on the original data, where it captures 71% of variability.

Finally, **Figure 3** displays in pseudocolor (with the MATLAB default colormap), the first PCA images from Eqs. (4) and (5). Panel 3a shows $\boldsymbol{Z}_1$, a weighted average of

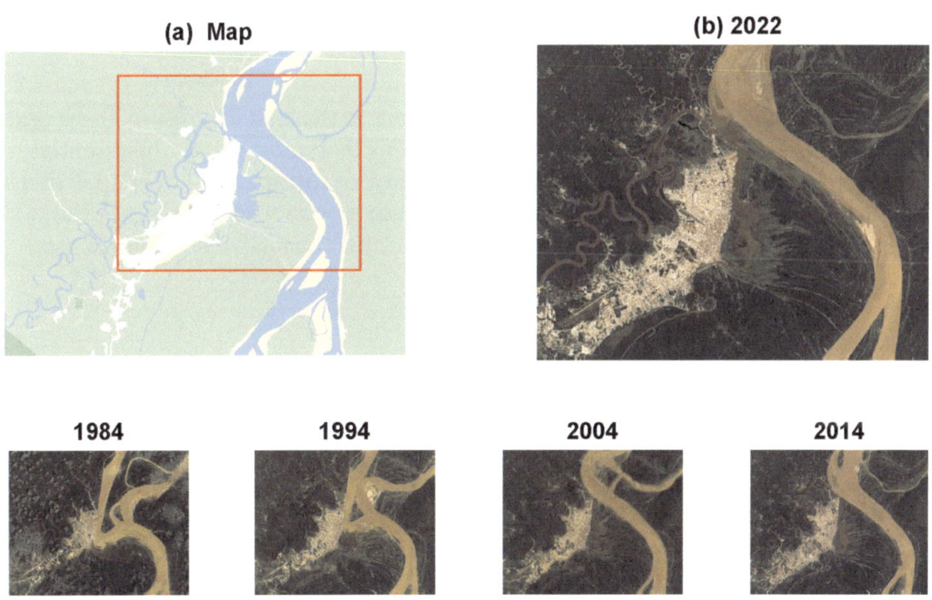

Figure 1.
Google Earth video [19] of Iquitos (Peru) in the period 1984–2022: (a) Google map 2023; (b) LandSat color image 2022; (c) Low-resolution decadal frames.

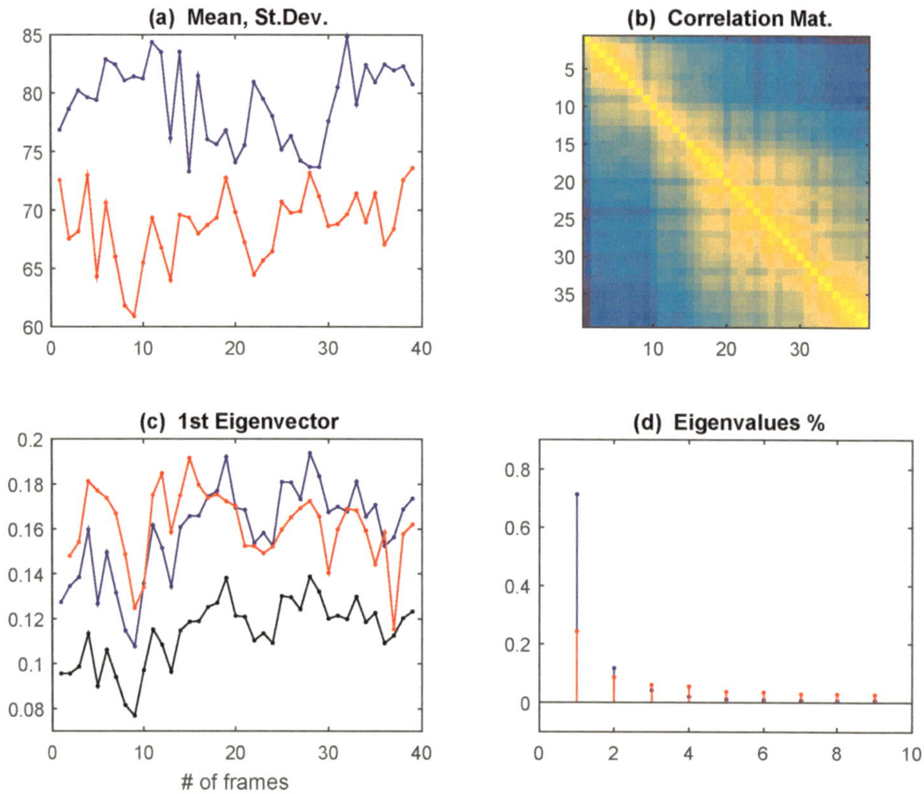

Figure 2.
Statistics of the Iquitos video [19] in grayscale: (a) Trends of mean (blue) and standard deviation of the frames; (b) Correlation matrix of Eq. (3); (c) Path of PC1 eigenvectors $\boldsymbol{v}_1$: original data (3) (blue), augmented data matrix (6) (black), and differenced series y (red); (d) Percent PC eigenvalues of the original and differenced (red) data.

the frames; its equalized version in Panel 3b enhances the land changes over 40 years. Panel 3c shows $\boldsymbol{Z}_0$, the classical PC1; it is more effective in capturing the essential changes, placed where two basic colors (blue/yellow of forest/other) overlap. Panel 3d displays the PC1 obtained on the differenced series; it enhances the edges of the various images. The attempt to reconstruct the original image from the latter has provided meaningless results.

3. Dynamic factor models for videos

PCA is an explorative and descriptive technique of data analysis, whose numerical results may sometimes be difficult to interpret, also in image processing. Space–time ACR may be an issue for PCA estimates, but it is an asset for modeling Y_{ijt} and using the models for filtering and forecasting. In particular, out-of-sample forecasting of video frames is an objective test-bench for checking the effectiveness of numerical methods. In this context, PCA results may be useful for improving the parsimony of classical regression models, with the application of *dynamic factor* techniques (e.g., [20, 21]) to space–time data.

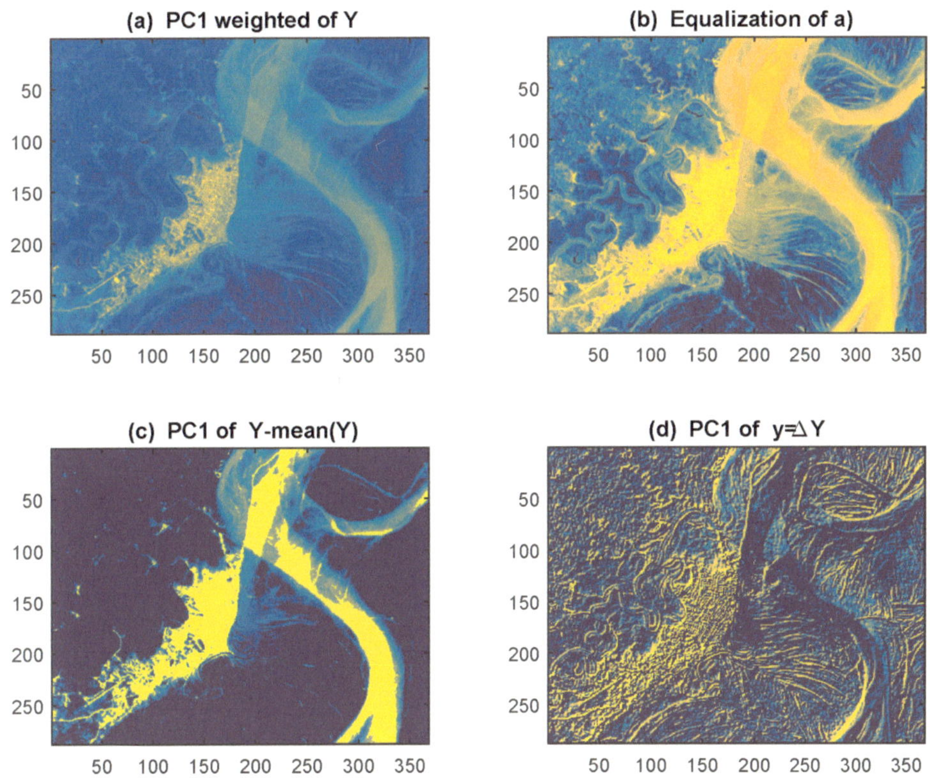

Figure 3.
Pseudocolor display of the first PCA images of Eqs. (4), (5): (a) Weighted average $\mathbf{Z}_1$; (b) Equalized version of $\mathbf{Z}_1$; (c) Centered image $\mathbf{Z}_0$; (d) Result on the differenced series.

3.1 Space-time systems

The space–time autoregressive (STAR) model puts each cell of the 3D array $\mathbb{Y}$ in relation to the contiguous ones, such as $Y_{ijt} = g\left(Y_{i\pm h, j\pm k, t-l}\right) + e_{ijt}$, where $h, k, l = 1, 2 \ldots p$ are spatial and temporal lags, $p > 0$ is the order of the dependence, and e_{ijt} is an unpredictable sequence. Under linearity of $g(\cdot)$, the STAR(p) representation can be denoted as

$$Y_{ijt} = \alpha_0 + \sum_{h=0}^{p} \sum_{k=0}^{p} \theta_{hk} Y_{i-h, j-k, t} + \sum_{l=1}^{p} \sum_{h=-p}^{p} \sum_{k=-p}^{p} \phi_{hkl} Y_{i-h, j-k, t-l} + e_{ijt}, \tag{7}$$

where $\theta_{00} = 0$ and $e_{ijt} \sim \text{IN}\left(0, \sigma_e^2\right)$ is an independent and normal (IN) sequence. The first part of the model (7), with parameters θ_{hk}, has a peculiar nature; it deals with the simultaneous relationships between the cells, and it is well studied in the (static) spatial AR literature, e.g., [18]. For reasons of identification and prediction, it has a *triangular* structure, so that filterings may proceed from the upper-left corner of every $\boldsymbol{Y}_t$ to the lower-right one in a sequential way.

Despite the simultaneous constraint, the number of parameters M_p of the model (7) is still large: for small p = 2 it becomes M_p = 59. A way to reduce the parametric complexity is to aggregate the pixels according to the geometric rules of contiguity of the chess moves. By defining the notation $ij-1$ when at least one of the indices i, j is lagged, then one may define the triangular (W) and ring (X) averages

$$W_{ij-1,t} = \left(Y_{i-1,j,t} + Y_{i,j-1,t} + Y_{i-1,j-1,t}\right)/3, \tag{8}$$

$$\begin{aligned} X_{ij-1,t-1} = (&Y_{i-1,j,t-1} + Y_{i+1,j,t-1} + Y_{i,j-1,t-1} + Y_{i,j-1,t-1} + \dots \\ &+Y_{i-1,j-1,t-1} + Y_{i+1,j-1,t-1} + Y_{i-1,j+1,t-1} + Y_{i+1,j+1,t-1})/8, \end{aligned} \tag{9}$$

which lead to a constrained STAR (2) model with only 9 coefficients

$$Y_{ijt} = \alpha_0 + \theta_1 W_{ij-1,t} + \theta_2 W_{ij-2,t} + \phi_1 Y_{ij,t-1} + \phi_2 Y_{ij,t-2} + \dots \tag{10}$$

$$+\beta_1 X_{ij-1,t-1} + \beta_2 X_{ij-2,t-1} + \beta_3 X_{ij-1,t-2} + \beta_4 X_{ij-2,t-2} + e_{ijt}, \tag{11}$$

The approach (8)–(11) of parametric reduction is subjective as depends on the aggregation rules; a more general solution arises by noting that all neighbors $\{Y_{i\pm h,j\pm k,t}\}$ of the series $\{Y_{ijt}\}$ can be "averaged" with the PCA technique and replaced by the first latent factor $\{Z_{ijt}\}$ as in Eqs. (4) and (5). Specifically, let $\boldsymbol{y}_{ij} = [Y_{ij1}, Y_{ij2} \dots Y_{ijT}]'$ be the time series located at ij, and consider its neighbors $\boldsymbol{y}_{i\pm h,j\pm k}$ in the square $\pm p$; then, for each ij one has the *temporal* PC1 component

$$\boldsymbol{z}_{ij} = \boldsymbol{\mathcal{Y}}_{ij}\boldsymbol{v}_{ij1}, \qquad \boldsymbol{\mathcal{Y}}_{ij} = \left[\boldsymbol{y}_{i-p,j-p} \cdots \boldsymbol{y}_{i-h,j+k} \cdots \boldsymbol{y}_{i+p,j+p}\right], \tag{12}$$

where the data matrices $\boldsymbol{\mathcal{Y}}_{ij}$ have size $T \times (2p+1)^2$–1.

Alternatively, with the $(2p+1)^2$–1 spatially lagged arrays implied by Eq. (12), one may proceed in the space domain and extract the PCs at each t. Specifically, let $\boldsymbol{Y}_t = \{Y_{ijt}\}$ be the t-th frame and $\boldsymbol{Y}_{hk,t} = \{Y_{i+h,j+k,t}\}$, $h, k = 1 \dots p$ its spatially shifted companions, i.e., the frames built with all h, k-lagged neighbors of each site ij; then, the *spatial* PC1 components are given by

$$\boldsymbol{z}_t = \boldsymbol{\mathcal{Y}}_t \boldsymbol{v}_{t1}, \qquad \boldsymbol{\mathcal{Y}}_t = \left[\mathrm{vec}\left(\boldsymbol{Y}_{-p,-p,t}\right) \dots \mathrm{vec}\left(\boldsymbol{Y}_{-h,+k,t}\right) \dots \mathrm{vec}\left(\boldsymbol{Y}_{ppt}\right)\right], \tag{13}$$

where the data matrices $\boldsymbol{\mathcal{Y}}_t$ have size $nm \times (2p+1)^2$–1, and reshaping $\boldsymbol{z}_t$ provides a $n \times m$ PC1 layer for each t.

Both approaches (12) and (13) yield latent factor arrays $\mathbb{Z}$ of size $n \times m \times T$, which may be used as explanatory (X) variables for the original data $\mathbb{Y}$, modeled as an ARX system. Further, the factor series may be represented by a simple AR scheme; this leads to the latent factor system of order $p = 2$ with 9 parameters, as Eqs. (10) and (11)

$$Z_{ijt} = \alpha_1 + \theta_1 Z_{ij,t-1} + \theta_2 Z_{ij,t-2} + u_{ijt}, \qquad u_{ijt} \sim \mathrm{IN}\left(0, \sigma_u^2\right), \tag{14}$$

$$Y_{ijt} = \alpha_2 + \phi_1 Y_{ij,t-1} + \phi_2 Y_{ij,t-2} + \beta_0 Z_{ijt} + \beta_1 Z_{ij,t-1} + \beta_2 Z_{ij,t-2} + e_{ijt}, \tag{15}$$

By means of the PCA framework (12) and (13), one can avoid to estimate the system (14) and (15) with the Kalman filter [21], which is nonlinear as regards factors and parameters. In the following, we campare the fitting and forecasting performaces of the models (10), (11), (14), and (15); this requires the definition of estimation algorithms.

3.2 Statistical algorithms

If the extraction of latent factors involves linear matrix algebra (eigenvalue decomposition), the estimation of parameters of models (10), (11), (14), and (15) may be accomplished with least squares (LS). Rewrite the model (10) and (11) in regression form as follows

$$Y_{ijt} = \boldsymbol{\delta}'\boldsymbol{\xi}_{ijt} + e_{ijt}, \tag{16}$$

$$\boldsymbol{\delta}' = [\alpha_0, \theta_1 \dots \phi_2, \beta_1 \dots \beta_4], \tag{17}$$

$$\boldsymbol{\xi}'_{ijt} = \left[1, W_{ij-1,t} \dots Y_{ij,t-2}, X_{ij-1,t-1} \dots X_{ij-2,t-2}\right], \tag{18}$$

where the regressors W, X are generated from the data $\{Y_{ijt}\}$ with the formulas (8), (9). Thus, minimizing the sum $\sum_{ijt} e^2_{ijt}(\boldsymbol{\delta})$ provides the LS estimator

$$\hat{\boldsymbol{\delta}}_N = \left(\sum_{t=p+1}^{T}\sum_{i=p+1}^{n-p}\sum_{j=p+1}^{m-p} \boldsymbol{\xi}_{ijt}\boldsymbol{\xi}'_{ijt}\right)^{-1} \sum_{t=p+1}^{T}\sum_{i=p+1}^{n-p}\sum_{j=p+1}^{m-p} \boldsymbol{\xi}_{ijt} Y_{ijt}, \tag{19}$$

$$= \boldsymbol{\delta} + \left(\sum_{t=p+1}^{T}\sum_{i=p+1}^{n-p}\sum_{j=p+1}^{m-p} \boldsymbol{\xi}_{ijt}\boldsymbol{\xi}'_{ijt}\right)^{-1} \sum_{t=p+1}^{T}\sum_{i=p+1}^{n-p}\sum_{j=p+1}^{m-p} \boldsymbol{\xi}_{ijt} e_{ijt}, \tag{20}$$

where $p = 2$ is the order of the model (7). Eq. (20) is obtained from Eqs. (16), (19), and $N = (n-2p)(m-2p)(T-p)$ is the actual sample size, net of the borders.

The expression (20) shows the consistency of the estimator (19), when the contemporaneous terms W_{ijt} have a triangular structure as (8); in fact, it enables the sequential calculation of residuals and their independence from all regressors: $\mathrm{E}\left(\boldsymbol{\xi}_{ijt} e_{ijt}\right) = \mathbf{0}$. Under this constraint and the conditions of stationarity and isotropy, one can show the classical convergence property [13].

$$\sqrt{N}\left(\hat{\boldsymbol{\delta}}_N - \boldsymbol{\delta}\right) \to \mathrm{N}\left[\mathbf{0}, \mathrm{E}\left(\boldsymbol{\xi}_{ijt}\boldsymbol{\xi}'_{ijt}\right)^{-1}\sigma^2_e\right], \quad \text{as} \quad N \to \infty, \tag{21}$$

from which, the dispersion matrix of the estimator (19), (20) is given by

$$\hat{\Sigma}_{\hat{\delta}} = \boldsymbol{R}_N^{-1}\hat{\sigma}^2_e, \qquad \boldsymbol{R}_N = \sum_{t=p+1}^{T}\sum_{i=p+1}^{n-p}\sum_{j=p+1}^{m-p} \boldsymbol{\xi}_{ijt}\boldsymbol{\xi}'_{ijt}, \tag{22}$$

$$\hat{\sigma}^2_e = \left(N - M_p\right)^{-1} \sum_{t=p+1}^{T}\sum_{i=p+1}^{n-p}\sum_{j=p+1}^{m-p} \left(Y_{ijt} - \hat{\boldsymbol{\delta}}'_N\boldsymbol{\xi}_{ijt}\right)^2, \tag{23}$$

where $M_p = 9$ is the length of $\boldsymbol{\delta}$ when $p = 2$. Unlike maximum likelihood, the algorithm (19) can manage datasets of large dimensions; i.e., with high pixel resolution $n \times m$ and high frequency frames T. Also the inversion of the $M_p \times M_p$ matrix (22) generally does not involve numerical issues. The formulas (19)–(23) can also be applied to the factor model (14), (15), in an even simpler way.

Forecasting. As regards prediction, given the linearity of Eqs. (10), (11) and (14), (15) the forecasts can be obtained with the chain rule of forecasting; i.e., by "pushing forward" the entries of the vector $\boldsymbol{\xi}_{ijt}$ and updating it with the past forecasts:

$$\hat{Y}_{ij,T+l} = \mathrm{E}\left(Y_{ij,T+l}|\boldsymbol{Y}_{T-t}, t \geq 0\right) = \boldsymbol{\delta}'\hat{\boldsymbol{\xi}}_{ij,T+l}, \tag{24}$$

$$\hat{\boldsymbol{\xi}}_{ij,T+l} = \left[1, \hat{W}_{ij-1,T+l} \dots \hat{Y}_{ij,T+l-2} \dots \hat{X}_{ij-2,T+l-2}\right]', \quad l = 1, 2 \dots L, \tag{25}$$

where $\hat{X}_{ij-2,T+l-2} = \left(\hat{Y}_{i-2,j-2,T+l-2} + \hat{Y}_{i-2,j-1,T+l-2} + \dots + \hat{Y}_{i+2,j-2,T+l-2}\right)/16$ are the lagged ring averages for $l > 2$, and so on.

A computational issue with the formula (24) and (25) applied to the model (10) and (11) arises from the presence of the contemporaneous terms $\hat{W}_{ij-k,T+l}, k = 1, 2$, which depend on the forecasts $\hat{Y}_{i-k,j-k,T+l}$ themselves. Although these elements satisfy the unilateral constraint, they require the upper-left border values $\hat{Y}_{ij,T+l}, i, j \leq p$ to start. These quantities can be obtained as forecasts of AR models applied to the marginal cells, or by simply setting $\hat{Y}_{ij,T+l} = Y_{ij,T}$, as in random-walk models. Hence, by the concatenation of forecasts in Eqs. (24) and (25), these marginal values influence the entire forecast frame $\hat{\boldsymbol{Y}}_{T+l}$. This effect can be checked empirically with out-of-sample forecasting, i.e., predicting data that have been omitted from parameter estimation. The typical statistics used for evaluation are the mean squared forecast errors (MSFE) and its relative R^2 index

$$\mathrm{MSFE}_{T'}(l) = \frac{1}{(n-2p)(m-2p)} \sum_{i=p+1}^{n-p} \sum_{j=p+1}^{m-p} \left(Y_{ij,T'+l} - \hat{Y}_{ij,T'+l}\right)^2, \tag{26}$$

$$R^2(l) = 1 - \mathrm{MSFE}_{T'+l}/\hat{\sigma}^2_{Y_{T'+l}}, \qquad l = 1, 2 \dots L, \tag{27}$$

where $T' = T - L$. These statistics enable to compare the performance of the models (10), (11) and (14), (15); the best one is that with lowest (26). The index (27) provides a measure of the reliability of forecasts in decision-making.

Simulations. To test the statistical properties of the LS estimator (19), we perform simulation experiments on the model (10), (11) with stable/unstable parameters and unilateral/multilateral simultaneous components. In particular, the multilateral design of the term W follows the cross (rook) scheme

$$W^*_{ij-1,t} = \left(Y_{i,j-1,t} + Y_{i,j+1,t} + Y_{i-1,j,t} + Y_{i+1,j,t}\right)/4,$$

in this case, the LS method should be biased for all parameters. We perform 500 replications of the system (10), (11) with $p = 1$, $n = 13$, $m = 12$, $T = 11$, and $|\delta_i| = 0.5$, 1.0, for $i = 1 \dots 4$; notice that the actual sample size $N = (n-2)(m-2)(T-1) = 1100$ is large enough. Simple means, root mean squared errors (RMSE = $[\mathrm{Var} + \mathrm{Bias}^2]^{1/2}$) and mean P-value of the normality test are reported in **Table 1**.

Since the parameters have the same size, their statistics can be averaged to provide synthetic indicators. As a result, one may note that the worst performance is that of multilateral W with "unstable" coefficients; whereas the model with unilateral W benefits from the greater signal-to-noise ratio σ_Y/σ_e induced by $|\delta_i| = 1$. This property is known as super-consistency [13], but its side-effect is the non-normality of estimates; this complicates statistical inference, requiring non-standard distributions as in the tests for unit-roots of time series.

		W-triangular			W-cross		
Param.	**Value**	**Mean**	**RMSE**	**N-test**	**Mean**	**RMSE**	**N-test**
α_0	0.5	0.501	0.031	0.414	0.211	0.291	0.055
θ_1	−0.5	−0.498	0.019	0.111	−0.433	0.072	> 0.5
ϕ_1	0.5	0.500	0.007	0.001	0.755	0.260	> 0.5
β_1	−0.5	−0.501	0.016	0.002	0.051	0.556	> 0.5
Abs. Ave.	0.5	0.500	0.018	0.132	0.363	0.295	0.389
α_0	1.0	1.001	0.030	> 0.5	−9.593	13.56	> 0.5
θ_1	−1.0	−1.000	0.000	0.001	−2.401	1.409	0.011
ϕ_1	1.0	1.000	0.000	0.001	1.086	0.107	0.001
β_1	−1.0	−1.000	0.000	0.001	2.707	3.709	0.001
Abs. Ave.	1.0	1.000	0.008	0.126	3.946	4.697	0.128

Table 1.
Performance of the LS estimator (19), applied to the model (10), (11) with order p=1, with triangular and cross (rook) contemporaneous component W, and stable (0.5) and unstable (1.0) coefficients; mean, RMSE, mean P-value are over 500 replications.

3.3 The empirical application

We illustrate and check the methods discussed so far on the same dataset as Section 2, displayed in **Figure 1**. The first step in model building is the extraction of the PC1 from the array $\mathbb{Y}$ with the approaches (12) and (13). With $p = 2$, we select 24 neighbor series of each site ij; the results are displayed in **Figure 4**. Panel (a) shows that the PC1 explains on average about 85% of the frame variability, with a growing trend; further, Panel (b) shows its spatial pattern, with a greater quote in areas subject to major changes. Panels (c,d) show that the PC1 arrays have an image representation that is sharp for the method (12) and blurred for (13).

Apart from the descriptive aspects, PC1 data are necessary for modeling and prediction; **Table 2** provides the parameter estimates of the models (10), (11), (14), (15) with the LS method (19). The estimations are performed on the frames $[\boldsymbol{Y}_3 \dots \boldsymbol{Y}_{T-2}]$, where $\boldsymbol{Y}_1, \boldsymbol{Y}_2$ are starting values and $\boldsymbol{Y}_{T-1}, \boldsymbol{Y}_T$ serve for forecasting evaluation. The results are very significant in terms of t-type statistics because these are inflated by the large sample size $N = 3{,}618{,}160$; the evaluation of coefficients should then be based on their size, e.g., $|\hat{\delta}_i| > 0.1$ as they are related to ACR. In any case, the R^2 indices of goodness of fit are very satisfactory, and, as regards the model (10), (11), the simultaneous components W have a leading role. However, the best fitting model is Eqs. (14) and (15) with factor component Z estimated as in Eq. (13), and normalized by $1/\|\boldsymbol{v}_1\|$; in the following, we evaluate their forecasting ability.

Forecasting. The prediction ability of the models of **Table 2** is evaluated on the last two frames $\boldsymbol{Y}_{T-1}, \boldsymbol{Y}_T$, which were kept out of parameter estimations. The forecasts were computed with the function (24), (25) and then evaluated with the statistic (26), (27), with starting point $T' = 2020$; in these computations, the forecasts were expressed in uint8 format, and the results are displayed in **Table 3**. As for the in-sample fitting, the best model is (14), (15) with factor (13) normalized by $1/\|\boldsymbol{v}_1\|$; it is

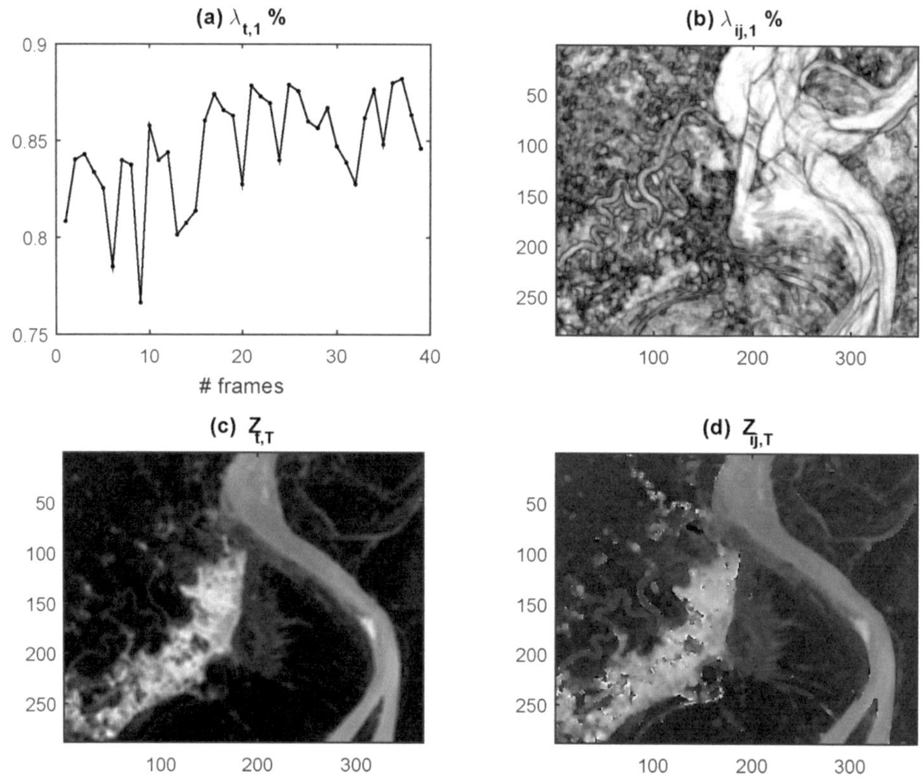

Figure 4.
Results of the extraction of the PC1 from the 24 neighbors of each ij: (a,c) Spatial direction (13); (b,d) Temporal direction (12); (a,b) Quote of variance explained by PC1; (c,d) Images of PC1 at time T = 2022.

Param.	Model (10)	(10) without W	Param.	Model (12), (15)	Model (13), (15)
α_0	0.804 (62.4)	2.958 (150.5)	α_1	3.284 (225.3)	4.768 (268.6)
θ_1	0.914 (1580)	.	θ_1	0.641 (1328)	0.651 (1330)
θ_2	−0.197 (−338.2)	.	θ_2	0.319 (662.)	0.293 (599.7)
ϕ_1	0.494 (641.1)	0.526 (445.5)	α_2	−0.436 (−36.6)	1.433 (102.3)
ϕ_2	0.204 (267.1)	0.217 (185.5)	ϕ_1	0.5252 (1048)	0.577 (1163)
β_1	−0.450 (−288.0)	0.019 (8.18)	ϕ_1	0.212 (426.9)	0.248 (503.5)
β_2	0.139 (117.0)	0.099 (56.4)	β_0	1.067 (2536)	0.722 (1812)
β_3	−0.185 (−121.2)	−0.006 (−2.53)	β_1	−0.565 (−790.5)	−0.408 (−701.1)
β_4	0.071 (62.5)	0.110 (63.2)	β_2	−0.233 (−347.3)	−0.159 (−298.1)
σ_e^2	90.39	212.05	σ_e^2	**77.60**	115.07
R^2	0.932	0.842	R^2	**0.942**	0.914

Table 2.
*LS estimates (and t-type statistics) (19), of the parameters of the model (10), (11) (with and without the contemporaneous components W); and model (14), (15) (with factors (13) and (12) expressed in uint8 format) on the data of **Figure 1** in the period 1986–2020.*

Year	(10)+W	(10)−W	(12), (15)	(13), (15)
2021	248.75	170.05	**166.47**	167.81
	(0.850)	(0.898)	(0.899)	(0.898)
2022	428.75	288.27	**281.74**	283.93
	(0.756)	(0.836)	(0.838)	(0.837)

Table 3.
MSFE statistics (17), and R^2 indexes (in parenthesis), of the AR models in **Table 2** *on the images 2021, 2022 of the Iquitos video; Bold indicates the best result.*

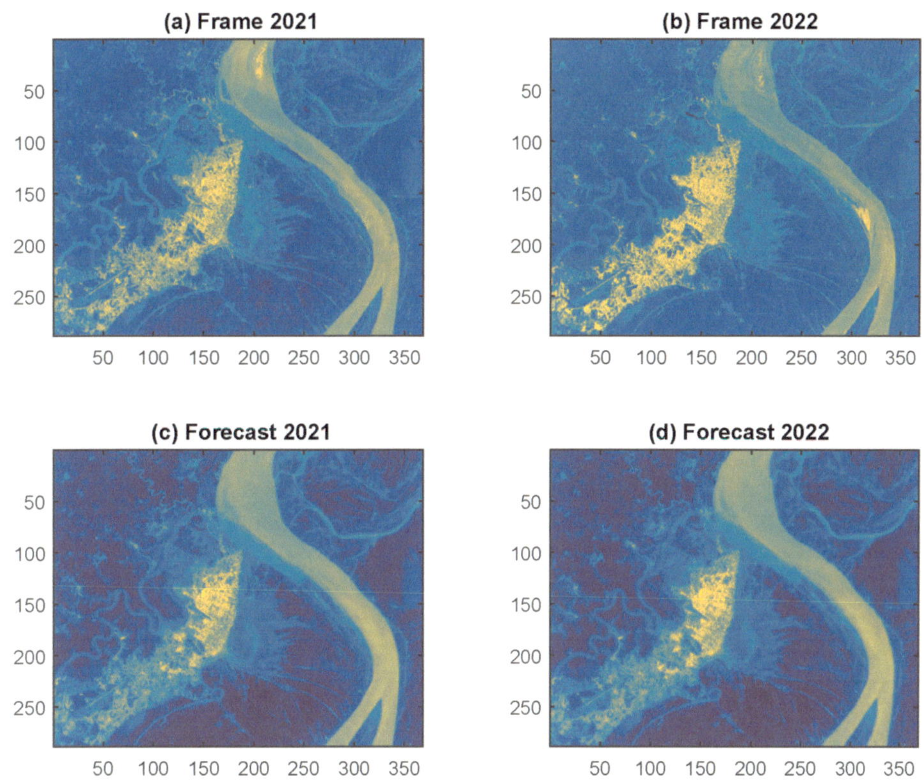

Figure 5.
Comparison of real frames and best forecasts in **Table 3**, *obtained with algorithm (24), (25): (a,b) Real images 2021, 2022; (c,d) Forecasts of the model (13)–(15).*

slightly better than model (14), (15) with factor (12) and model (10), (11) without the simultaneous components. In general, while the W components improve fitting, they may affect forecasts; the reason is partly due to the starting values in the (upper-left) borders, which are established with the random walk rule, e.g., $\hat{Y}_{11,T+1} = Y_{11,T}$.

Finally, **Figure 5** displays the best forecasts for 2021, 2022 (with smallest MSFE in **Table 3**) and compares them with the ground images (in pseudocolor MATLAB). The spatial paths of forecasts are consistent with the actual images, although they show fewer details in the urban area; further, the $R^2 = 0.9$ coefficient in **Table 3** confirms the reliability of these forecasts.

4. Conclusions

This paper is concerned with the application of multivariate statistical methods to timelapse videos of remotely sensed data. Such movies are nowadays available on Internet for entertaining and scientific purposes, and their modeling is challenging both for the size of datasets (big data) and the complexity of the phenomena they represent. Classical multivariate techniques of data reduction, such as principal components, are useful both for reasons of data description and model building. In particular, PCA may condense the frames as a long exposure photography (see **Figure 3**) and provide local components for dynamic factor models, as Eqs. (14), (15). These are smart alternatives to complex space–time AR models.

The PCA of videos in the time domain provides two basic solutions for frame fusion, depending on whether it considers a weighted average of original images or the centered data array, see Eqs. (4), (5). In the second case, a uint8 transformation of the estimates is necessary to appreciate the result as an image; this transformation, by censoring negative values, makes the final result more essential and highlights major land changes (see **Figure 3c**). However, also factor models are useful for descriptive purposes as highlights the spectral properties of data arrays in the time domain and the spatial domain (see **Figure 4a,b**).

Mathematical modeling is useful for out-of-sample forecasting; this, in turn, is useful for monitoring and surveillance. The paper has provided a factor model framework which is more parsimonious and effective than the classical STAR systems. In prediction, there is usually a negative trade-off between model complexity (which improves fitting) and out-of-sample performance. While the model (7) requires ad-hoc aggregations of neighboring pixels, as in Eqs. (8), (9), in the model (14), (15) one has only to select the time order. As regards the spatial dimension, one may increase the span p in Eqs. (12) and (13) without increasing the number of coefficients of (14), (15); the only drawback is PC estimation at the borders of the lattice which requires symmetrical or circulant integrations.

A comparison of the models in **Table 3** with $p = 2$ shows that there is *not* significant difference in the performance of factor models and simplified STAR (without the simultaneous component W). In particular, all R^2 coefficients are close to 90%; however, the distance between the two modelings may increase as p.

Data and software

They are available at the site: https://it.mathworks.com/matlabcentral/fileexchange/173895-pca-and-factor-ar-models-for-timelapse-video-data

Author details

Carlo Grillenzoni
IUAV University, Venice, Italy

*Address all correspondence to: carlog@iuav.it

References

[1] Google Earth. Google Earth Engine. Mountain View (CA): Google Earth; 2024. Available from: https://earthengine.google.com/timelapse [Accessed: June 30, 2024]

[2] Mudrová M, Procházka A. Principal Component Analysis in Image Processing. 2005. Available from: https://www2.humusoft.cz/www/papers/tcp05/mudrova.pdf [Accessed: June 30, 2024]

[3] Pandey PK, Singh Y, Tripathi S. Image processing using principle component analysis. International Journal of Computer Applications. 2011; **15**(4):37-40

[4] Salata S, Grillenzoni C. A spatial evaluation of multifunctional ecosystem service networks using principal component analysis: A case of study in Turin, Italy. Ecological Indicators. 2021; **127**:107758

[5] Grillenzoni C. Sequential mean shift algorithms for space–time point data. Environmental Earth Sciences. 2018;**77**: 336

[6] Stahlschmidt S, Härdle WK, Thome H. An application of principal component analysis on multivariate time-stationary spatio-temporal data. Spatial Economic Analysis. 2015;**10**(2): 160-180

[7] Krzyśko M, Nijkamp P, Ratajczak W, Wolyński W, Wenerska B. Spatio-temporal principal component analysis. Spatial Economic Analysis. 2024;**19**(1): 8-29

[8] Zamprogno B, Reisen VA, Bondon P, Aranda-Cotta HHC, Reis NC. Principal component analysis with autocorrelated data. Journal of Statistical Computation and Simulation. 2020;**90**(12):2117-2135

[9] Sahouria E, Zakhor A. Content analysis of video using principal components. IEEE Transactions on Circuits and Systems for Video Technology. 1999;**9**(8):1290-1298

[10] Liu Y, Liu Y, Chan KCC. Dimensionality reduction for descriptor generation in rushes editing. In: IEEE International Conference on Semantic Computing, Santa Monica (CA). New York (NY); 2008. pp. 104-111

[11] Yousif AJ, Al-Jammas MH. Exploring deep learning approaches for video captioning: A comprehensive review. e-Prime. 2023;**6**:100372

[12] Weissenborn D, Täckström O, Uszkoreit J. Scaling autoregressive video models. In: International Conference on Learning Representations (ICLR 2020). ICLR 2020 Conference. 2020. Available from: https://iclr.cc/Conferences/2020

[13] Grillenzoni C. Adaptive spatio-temporal models for satellite ecological data. Journal of Agricultural, Biological and Environmental Statistics. 2004;**9**: 158-180

[14] Chang R. Application of principal component analysis in image signal processing. In: Proceedings International Conference on Image, Signal Processing and Pattern Recognition (ISPP 2022). Vol. 12247. SPIE Digital Library; 2022. Available from: https://2022.icispp.com/

[15] Jolliffe IT. Principal Component Analysis. New York, NY: Springer; 2002

[16] Vanhatalo E, Kulahci M. Impact of autocorrelation on principal components

and their use in statistical process control. Quality and Reliability Engineering International. 2015;**32**(4): 1483-1500

[17] Jombart T, Devillard S, Dufour AB, Pontier D. Revealing cryptic spatial patterns in genetic variability by a new multivariate method. Heredity. 2008; **101**(1):92-103

[18] Grillenzoni C. Forecasting lattice and point spatial data: Comparison of unilateral and multilateral SAR models. Forecast. 2024;**6**(3):700-717

[19] Google Earth. Iquitos, Peru - Earth Timelapse. Mountain View (CA): Google Earth; 2023. Available from: https://www.youtube.com/watch?v=ZHhByopdLY4 [Accessed: June 30, 2024]

[20] Stock JH, Watson MW. Dynamic factor models, factor-augmented vector autoregressions, and structural vector autoregressions in macroeconomics. In: Handbook of Macroeconomics. Vol. 2. Amsterdam: Elsevier; 2016. pp. 415-525

[21] Krantz S. Dynamic Factor Models: A Very Short Introduction. Cran. R Project. Virtual Publisher; 2023. DOI: 10.32614/CRAN.package.dfms. Available from: https://cran.r-project.org/

Chapter 9

Small-Signal Stability Analysis of Virtual Impedance Based Parallel Inverters

Deependra Neupane and Nawaraj Poudel

Abstract

Maintaining voltage-frequency levels and balancing active and reactive power are significant issues in parallel inverters used in standalone electrical systems. Therefore, modelling and analysing parallel inverters has become crucial in developing appropriate control mechanisms to mitigate power-sharing and voltage-frequency issues in these systems. This study has developed a non-linear model of a parallel inverter consisting of a conventional Proportional Integral (PI) control with virtual impedance control. The equation system was linearised for zero initial conditions and steady-state conditions. The effect of load changes on the system was also studied. Additionally, eigenvalue and frequency response analyses were performed. The results of the study show that the developed non-linear model matches the electrical simulation performed in Simulink. The frequency dynamics and voltage profile of the system were improved by using virtual impedance control. The system's damping increased by 2% for the most dominant pole, and the reactive power sharing also increased.

Keywords: parallel inverter, small signal modelling, virtual impedance, microgrid, stability analysis

1. Introduction

With the growth of distributed generation systems and loads, it has become more critical to maintain an uninterruptible power supply system. One of the several solutions to obtain a continuous supply to cope with growing demand is to use parallel inverters. However, appropriate control mechanisms have to be ensured in individual inverters for proper operation, load sharing and stability. One of the significant challenges in parallel systems is to minimise the circulating current between inverters, maintain the voltage and frequency level and balance the active and reactive power between sources and loads [1, 2]. Less circulating current can only be obtained by increasing the output impedance of the inverters. To accomplish this, virtual impedance can be added to the inverter's control loop. Due to large output inductor values or distances between the units, inverters' line and output impedances are typically primarily inductive. However, this is only sometimes the case, as system characteristics and the chosen control method affect the inverter output impedance. Several applications of virtual impedance have been shown in parallel inverter systems, including

IntechOpen

harmonic voltage correction, better power sharing and increased stability [3]. Physical output impedance mismatches between inverters can typically be compensated using virtual impedances. The primary mechanism is to add additional phase displacement equivalent to line reactance to nullify the line impedance effect [4]. Similarly, the converter controller has effectively used the droop control approach for voltage and frequency balance and the proper active and reactive power sharing in parallel inverters [5, 6]. One significant advantage of this method is that it is widespread in low-voltage cable grids, where the line impedance is typically resistive [6]. However, it achieves current-sharing accuracy at the penalty of poor output voltage regulation. Several studies have been done to fix the inverter's output impedance to ensure power-sharing stability.

Early studies have proposed a control strategy that would apply to both linear and non-linear loads using the idea of the conventional frequency/voltage droop technique including the survey done by Tuldhar et al. [7–10]. The small signal modelling of the system was developed to study the dynamics of inverter systems in a microgrid. Pogaku has developed a detailed model of a microgrid considering current and voltage loops with droop power coefficient [11]. Zhang et al. presented a small signal model of two inverters in parallel without considering line impedance dynamics. Both of the open-loop and closed-loop analysis have been performed [12]. Small modelling and analysis of parallel inverters have also been developed [13, 14]. The developed model accuracy is tested by comparing the model solution with a simulation consisting detailed of switching model. A similar study has also done by Bennia [15] where the development is compared with the actual switching model using simpower system. Additionally, a small-signal model for an inverter under droop control operation is developed in this paper. An averaged model simulation using PLECS was used to determine stable gains for the system [16, 17]. Rasheduzzaman et al. have developed a reduced-order small signal model of a microgrid system for grid-tied/stand-alone systems by removing the voltage loop control [18]. A study by Braynt has presented a small signal model of a feeding inverter considering the modulator transport delay and performed stability studies [19]. Moreover, small signal stability assessments using singular perturbation with eigenvalue analysis for microgrid systems have been systematically studied by Mariani et al. [20]. A study by Sharma et al. presented a virtual impedance-based phase-locked loop (PLL) inverter control scheme in an islanded microgrid environment without implementing the conventional droop control strategy [21]. Several others are related to developing the small signal modelling of parallel inverters [22–31].

Along with the conventional droop control that includes droop control, voltage and current control loop, the use of a virtual impedance loop compensates for the additional voltage and phase caused due to line impedance. Studies on the effect of the use of virtual impedance in the control of parallel inverters have come a long way. Early studies performed by Refs. [6, 32] have proposed a control method allowing parallel inverters to operate in standalone/grid-connected mode by taking the lines X and R into account [6]. Furthermore, the study performed by Alcala et al. explored the effect of line impedance on power sharing in parallel inverters. The study showed improved reactive power sharing when virtual impedance control is implemented [1]. Zhong and friends have proposed a robust droop controller to provide proportional load sharing in parallel inverters because individual inverters per unit impedance are challenging to achieve [33].

Existing analysis techniques primarily include eigenvalue analysis utilising state space models, impedance analysis based on impedance models and other non-linear analysis techniques for the small signal stability of an inverter-based microgrid [34]. Several studies focused on the study's time domain simulation and eigenvalue analysis. However, a detailed description of linear and non-linear models with their appropriate validation must be included. The inductive load model included in several works of literature is represented as a first-order differential equation that can be expressed in the algebraic equation to reduce the computation time.

This study examines the application of virtual impedance-based controllers in parallel inverters for better system stability (both frequency and voltage). Much literature has been studied on the application of virtual impedance in controllers; however, a detailed time domain analysis and comparison with actual systems still need to be included. In light of the prior research gaps, the following contributions are intended to be presented by this study:

1. Developing a non-linear system of equations of the parallel inverters in terms of Differential-Algebraic Equations.

2. Comparison of the non-linear model with that of the result from the simulation result obtained from Simulink that uses elementary blocks.

3. To provide insight into the effect of virtual impedance on frequency and voltage dynamics of parallel connect inverter.

4. Simplified load model for a standalone system.

The work focuses on modelling various inverter components' filter and coupling impedances mathematically. The main components are the power component (inverter circuit), the control component (inverter operation control dynamics), the filters, and the load. The system of equations used to describe the behaviour of each element is derived from existing literature. Because they represent the traditional DQ control technique, the equations governing the dynamics of the inverter's control component are non-linear. As a result, the equations are linearised using Python's SYMPY module. The parameters derived from the linearised version of the equations are then evaluated on the original set of equations to validate the model and findings. Simple construction blocks are used in the Simulink system to validate the outcomes. The results from both scenarios are compared. In summary, this study aims to present the detailed mathematical formulation of virtual impedance-based parallel inverters, including both the power and control parts. However, this study has yet to consider the phase lock loop (PLL) dynamics for synchronising the inverters in between.

The remaining sections of the document are arranged as follows: Section 2 presents the mathematical modelling of each component of an independent parallel inverter. This section covers linearisation and eigenvalue computation. Sections 3 and 4 contain the simulation results, discussions and conclusions.

2. Modelling of components

The diagram in **Figure 1** shows the components of VSI. The circuits for power measurement and computation, the voltage and current controller, the droop controller and the inverter power circuit are the major components. A digital signal processor (DSP), voltage source inverter (VSI), is the VSI's control element. It

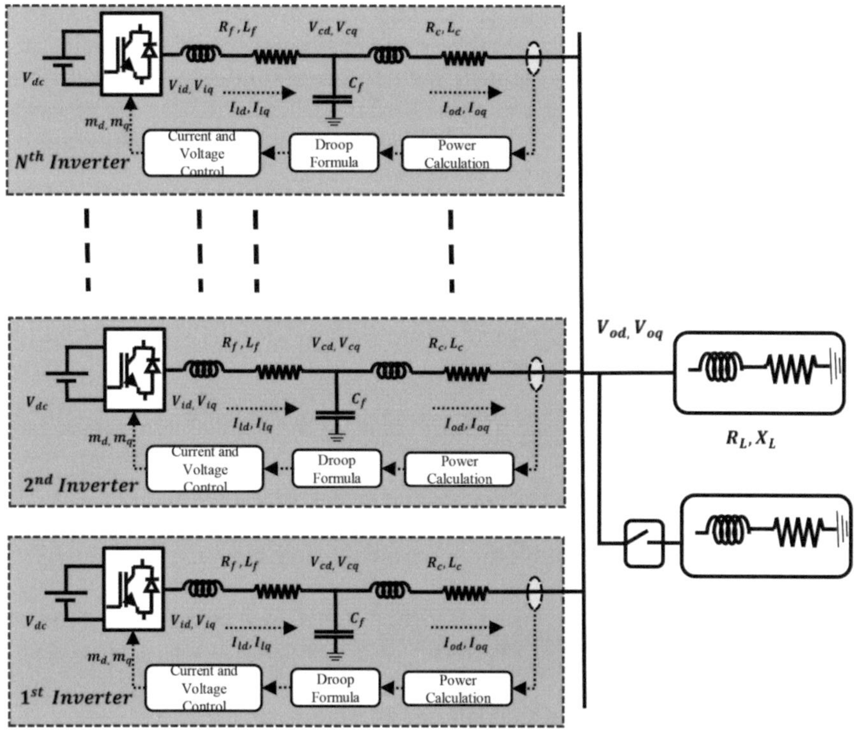

Figure 1.
Parallel inverters connected to load.

determines the instantaneous load voltage and filter current as well as the duty cycle of the inverter pulse-generating circuit. The details of the modelling for each component are explained below.

In microgrid applications, the direct quadrature (*DQ*) frame approach is the most popular method for small-scale signal modelling. It transforms the three-phase sinusoidal system model into a DC system model using a rotating DQ reference frame. This solution perfectly bridges the gap between the AC microgrid system with just a sinusoidal stable point and small signal models based on the continuous equilibrium point. Microgrids lack a single model because they differ in kind, topologies, architectures and control loops based on their control objectives [35].

The modelling approach presented in this study breaks down the entire system into two main sub-modules: inverter and loads. Each inverter is modelled based on its reference frame, which its local controller sets. On the other hand, the state equation of the loads is described using the reference frame of a specific inverter, also known as the typical reference frame. The variables in the instantaneous frame of reference are $[d, q]$, while the relative frame of reference is $[D, Q]$. The transformation matrix for converting between the individual frame and the standard frame is provided below:

$$[V_{D,Q}] = [T_i][V_{d,q}] \tag{1}$$

where $[T_i] = \begin{bmatrix} cos\,(\delta_i) & -sin\,(\delta_i) \\ sin\,(\delta_i) & cos\,(\delta_i) \end{bmatrix}$ the δ_i is the angle of the reference frame of *ith* inverter concerning the common frame of reference.

2.1 VSI inverter

2.1.1 Inverter with filter model

The coupling impedance and filter have the most effects on the inverter's dynamics. A set of differential equations in the dq0 frame can be used to represent the LC filter, as has been noted in the literature. The inverter output is another source of the input voltage used in the LC filter. The output voltage of the inverter in the dq0 frame of reference is given by a particular equation. **Figure 1** displays the general schematic of the load-connected inverter system with LC filter and coupling impedance.

Using the controlled duty cycle, the inverter generates v_{id} and v_{iq}, the dq voltage output. A first-order system with a matching time delay between the duty cycle and output voltage represents the relationship between the inverter's duty cycle and output voltage, according to the equation. The switching and snubber circuit dynamics of the inverter should be the focus of the small signal stability analysis. Nonetheless, a comprehensive model for the inverter and filter circuit has also been created in order to test the parameters using MATLAB/Simulink.

The actual physical VSI is made up of switching devices in a switching bridge circuit. However, the control signal given in the gate drivers of switching devices represents the inverter as a voltage source due to the intricacy of the modelling. The LC filter and the microgrid with coupling impedance are linked to the inverter's output. The filter for the single inverter unit is displayed in **Figure 1**. The following is an expression of the dynamics:

$$
\begin{aligned}
\frac{i_{ldi}}{dt} &= (-R_f/L_f)i_{ldi} + \omega_i i_{lqi} + (v_{idi} - v_{cdi})/L_f \\
\frac{i_{lqi}}{dt} &= -\omega_i i_{ldi} + (-R_f/L_f)i_{lqi} + (v_{iqi} - v_{cqi})/L_f \\
\frac{v_{odi}}{dt} &= \omega_i v_{oqi} + (i_{ldi} - i_{odi})/C_f \\
\frac{v_{oqi}}{dt} &= -\omega v_{odi} + (i_{lqi} - i_{oqi})/C_f \\
\frac{i_{odi}}{dt} &= (-R_{ci}/L_{ci})i_{odi} + \omega_i i_{oqi} + (v_{od} - v_{bd})/L_{ci} \\
\frac{i_{oqi}}{dt} &= -\omega_i i_{odi} + (-R_{ci}/L_{ci})i_{oqi} + (v_{oq} - v_{bq})/L_{ci}
\end{aligned}
\tag{2}
$$

The filter capacitance is denoted by C_f, the coupling resistance and inductance by R_c, L_c, and the filter resistance and inductance by R_f, L_f. Combining linear circuits with non-linear switching devices creates the inverter. However, system analysis becomes more difficult due to the mathematical model of individual switching circuits. The switching portion of the inverter's approximation has been developed in this work using the dq equivalent form. The duty cycle is used as the input in each frame to resolve the output from the inverter into the d and q frames. Equations provide the inverter's output.

The modulation index supplied to the inverter determines the inverter output voltage, which is represented by v_{idi} and v_{iqi} in dq coordinate. The instantaneous average value of the converter output voltage, ignoring the delay impact of PWN implementation, is determined by multiplying the modulation index by the real DC voltage. The per unit output converter voltage will therefore be almost equivalent to the voltage reference from the current controller if the modulation index is calculated using the division shown in the figure, as summarised by the following [36].

$$[m_d, m_q] = \frac{[v_{ld,ref}, v_{lq,ref}]}{V_{dc}}$$
$$[v_{id}, v_{iq}] \approx [m_d, m_q] V_{dc}$$

Hence,

$$[v_{id}, v_{iq}] = [v_{ld,ref}, v_{lq,ref}]$$

2.1.2 Droop control model

The instantaneous active and reactive power components are calculated from the measured output voltage and output current, as the following equation.

$$\begin{aligned} p_{inv,i} &= \frac{3}{2} \times (v_{od,i} i_{od,i} + v_{oq,i} i_{oq,i}) \\ q_{inv,i} &= \frac{3}{2} \times (v_{oq,i} i_{od,i} + v_{od,i} i_{oq,i}) \end{aligned} \tag{3}$$

In a standalone microgrid system, an individual inverter unit first determines its output power and then uses the droop characteristic to obtain the reference value of the output voltage's frequency and amplitude. After that, it adjusts its output to achieve a proper distribution of the active and reactive power of the entire microgrid power system. The inverter unit satisfies the droop relationship between reactive power and voltage amplitude and the droop relationship between active power and frequency. This results in proportional load sharing between inverters in the microgrid. Power is calculated from direct measurement of current, and power consists of noise; hence, an appropriate low pass filter is implemented to obtain a smooth control performance.

$$\begin{aligned} \frac{P_{inv,i}}{dt} &= -\omega_f P_{inv,i} + \omega_f p_{inv,i} \\ \frac{Q_{inv,i}}{dt} &= -\omega_f q_{inv,i} + \omega_f q_{inv,i} \end{aligned} \tag{4}$$

where $p_{inv,i}, q_{inv,i}$ are the active and reactive power calculated from Eq. (3) and $P_{inv,i}, Q_{inv,i}$ filter output of the measured active and reactive power. Similarly, the droop characteristics can be implemented as:

$$\begin{aligned} \omega_i &= \omega_0 - P_{invi} \times m_i \\ v_{od,ref,i} &= v_{d,ref} - Q_{inv,i} n_i \\ v_{oq,ref,i} &= v_{q,ref} \end{aligned} \tag{5}$$

The power angle of the ith inverter concerning the reference inverter can be written as:

$$\frac{d\delta_i}{dt} = w_i - w_{iref} \tag{6}$$

2.1.3 Virtual impedance model

In the context of virtual impedance-based control, an outer virtual impedance Zvo (s) is used to modify the reference of the AC current/voltage controller through feedback impedance $R_v + j\omega L_v$. This virtual impedance is subject to the dynamics of AC current/voltage control loops [37]. The study involves multiplying the output current of an inverter with a complex virtual impedance to change the reference input for the voltage controller. An equation gives the new reference input for the voltage controller.

$$\begin{aligned} v_{od,ref,i} &= v_{od,ref,i} - i_{odi}R_{vi} + i_{oqi}\omega_i L_{vi} \\ v_{oq,ref,i} &= v_{oq,ref,i} - i_{oqi}R_{vi} - i_{odi}\omega_i L_{vi} \end{aligned} \tag{7}$$

Virtual impedance will be written as $Z_{ci} = R_{vi} + j\omega_i L_{vi}$.

2.1.4 Voltage controller

The differential-algebraic equation system for the voltage controller as mentioned in **Figure 2** is given by Eq. (8). The standard Proportional Integral (PI) controller has been used to control the output voltage of VSI.

$$\begin{aligned} i_{ld,refi} &= K_{pv}\left(v_{odrefi} - v_{odi}\right) + K_{iv}\gamma_{di} + i_{odi} - \omega_i C_f v_{oqi} \\ i_{lq,refi} &= K_{pv}\left(v_{oqrefi} - v_{oqi}\right) + K_{iv}\gamma_{qi} + i_{oqi} - \omega_i C_f v_{odi} \\ \frac{\gamma_{di}}{dt} &= v_{odrefi} - v_{odi} \\ \frac{\gamma_{q1}}{dt} &= v_{oqrefi} - v_{oqi} \end{aligned} \tag{8}$$

2.1.5 Current controller

The system of differential-algebraic equation for voltage controller as mentioned in **Figure 2** is given by Eq. (9). The standard Proportional Integral (PI) controller has been used to control the output current of VSI.

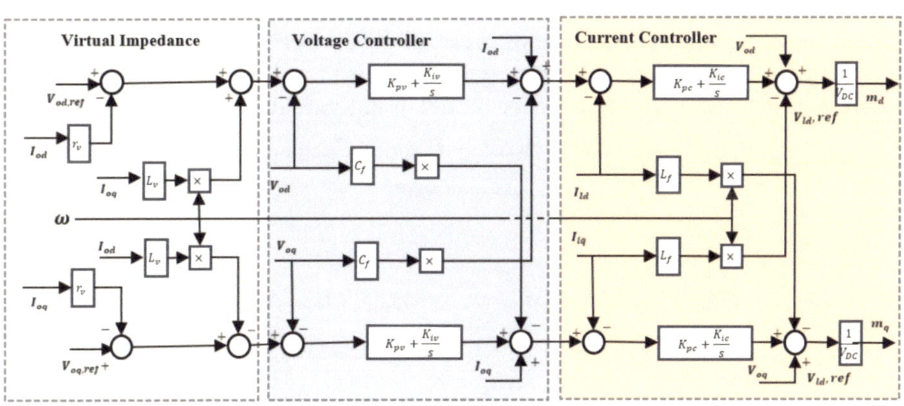

Figure 2.
Conventional PI with virtual impedance control strategy for an inverter adapted from Ref. [36].

$$\begin{aligned} v_{ld,refi} &= K_{pc}\left(i_{ldrefi} - i_{ldi}\right) + K_{ic}\zeta_{di} + v_{odi} - \omega_i L_f i_{lqi} \\ v_{lq,refi} &= K_{pc}\left(i_{lqrefi} - i_{lqi}\right) + K_{ic}\zeta_{qi} + v_{oqi} - \omega_i L_f i_{ldi} \\ \frac{\zeta_{d1}}{dt} &= if_{drefi} - i_{ldi} \\ \frac{\zeta_{q1}}{dt} &= if_{qrefi} - i_{lqi} \end{aligned} \tag{9}$$

2.2 Load modelling

Static loads are taken into account by the combination of resistor and inductor (RL load). In order to study the dynamic behaviour of the system, a load perturbation is applied parallel to the static load connected to the common bus, as shown in **Figure 1**. As stated by Eq. (10), the algebraic equation in the dq0 frame represents the generalised equations representing the load dynamic.

$$\begin{aligned} v_{oD} &= R_L \sum_{i=1}^{N} i_{oDi} - X_L \sum_{i=1}^{N} i_{oQi} \\ v_{oQ} &= R_L \sum_{i=1}^{N} i_{oQi} + X_L \sum_{i=1}^{N} i_{oDi} \end{aligned} \tag{10}$$

where R_L and X_L are the load resistance and inductive reactance of the load. The values of R_L and X_L are calculated based on load active and reactive power, P_L and Q_L respectively. The load resistance and reactance at full loading are computed as follows:

$$\begin{aligned} I_{nom} &= \frac{S_{base}}{\sqrt{3}Vnom} \\ R_L &= \frac{P_L}{3I_{nom}^2} \\ X_L &= \frac{Q_L}{3I_{nom}^2} \end{aligned} \tag{11}$$

3. Linearisation and state space model

Small signal stability investigations invariably play a crucial role in the design of power systems. Eigenvalues analysis is generally used to depict the SSS of power systems systematically. The state space model of individual inverter can be written as:

$$\begin{aligned} \dot{\Delta x_i} &= A_i \Delta x_i + B_i \Delta u_i + F_i \Delta v_i \\ \Delta y_i &= C_i \Delta x_i \end{aligned}$$

where,

$$\Delta x_i = \left[\Delta P_{inv,i}\ \Delta Q_{inv,i}\ \Delta\delta_i\ \Delta i_{ld,i}\ \Delta i_{lq,i}\ \Delta v_{od,i}\ \Delta v_{oq,i}\ \Delta i_{od,i}\ \Delta i_{oq,i}\ \Delta\gamma_{d,i}\ \Delta\gamma_{q,i}\ \Delta\zeta_{d,i}\ \ \Delta\ \zeta_{q,i}\right]$$

$$\begin{aligned} \Delta u_i &= \left[\ \Delta v_{bd} \quad \Delta v_{bq}\ \right] = \left[\Delta v_{b,dq}\right] \\ \Delta y_i &= \left[\ \Delta i_{od,i} \quad \Delta i_{oq,i} = \left[\Delta i_{o,dq}\right]\right] \\ \Delta v_i &= \left[\ \Delta v_{d,ref} \quad \Delta v_{q,ref} \quad \Delta\omega_0\ \right] \end{aligned}$$

$$\left[\Delta v_{b,dq}\right] = \left[T_s^{-1}\right]\left[\Delta v_{b,DQ}\right] + \left[T_v^{-1}\right]\left[\Delta\delta_i\right] \tag{12}$$

For load:

$$\Delta v_{b,DQ} = A_L \sum_{k=1}^{N} \Delta i_{ok,DQ} \tag{13}$$

where N is the total number of inverters feeding the load.

$$A_L = \begin{bmatrix} R_L & X_L \\ -R_L & X_L \end{bmatrix} \tag{14}$$

The $\Delta v_{b,DQ}$ is for a common frame of reference; for individual inverter frames, $\Delta v_{b,dq}$ have to be determined by using the transformation matrix:

$$\left[\Delta i_{o,DQ}\right] = [T_s]\left[\Delta i_{o,dq}\right] + [T_c][\Delta \delta_i] \tag{15}$$

A general state space equation for a single inverter can be derived as:

$$\Delta u_i = \left[\Delta v_{b,dq}\right] = \left[T_s^{-1}\right]\left[\Delta v_{b,DQ}\right] + \left[T_v^{-1}\right][\Delta \delta_i]$$

$$\Delta u_i = \left[T_s^{-1}\right][A_L] \sum_{i=1}^{N} \left[\Delta i_{io,DQ}\right] + \left[T_v^{-1}\right][\Delta \delta_i]$$

$$\Delta u_i = \left[T_s^{-1}\right][A_L][T_s] \sum_{i=1}^{N} \left[\Delta i_{oi,dq}\right] + \left[T_s^{-1}\right][A_L] \sum_{i=1}^{N} [T_{ci}][\Delta \delta_i] + \left[T_v^{-1}\right][\Delta \delta_i]$$

We get

$$\Delta u_i = \left[T_s^{-1}\right][A_L][T_s][C_i] \sum_{i=1}^{N} [\Delta x_i] + \left(\left[T_s^{-1}\right][A_L] \sum_{i=1}^{N} [T_{ci}] + \left[T_v^{-1}\right]\right)[A_{\delta i}][\Delta x_i]$$

Hence,

$$\Delta u_i = \left[A_i'\right] \sum_{i=1}^{N} [\Delta x_i] + \left[B_i'\right][\Delta x_i]$$

Hence, the state space model can be reformulated as:

$$\Delta \dot{x}_i = [A_i]\Delta x_i + [B_i]\left(\left[A_i'\right] \sum_{i=1}^{N} [\Delta x_i] + \left[B_i'\right][\Delta x_i]\right) + [F_i][\Delta v_i]$$

$$\left[\Delta \dot{x}_i\right] = \left([A_i] + [B_i]\left[B_i'\right]\right)[\Delta x_i] + [B_i]\left[A_i'\right][I_N][\Delta x_i] + [F_i][\Delta v_i]$$

$$\left[\Delta \dot{x}_i\right] = \left([A_i] + [B_i]\left[B_i'\right] + [B_i]\left[A_i'\right][I_N]\right)[\Delta x_i] + [F_i][\Delta v_i]$$

$$\Delta \dot{x}_i = \left[A''_i\right][\Delta x_i] + [F_i][\Delta v_i]$$

where $A''_i = [A_i] + [B_i]\left[B_i'\right] + [B_i]\left[A_i'\right][I_N]$
For n^{th} inverter system:

$$\begin{bmatrix} \Delta\dot{x}_1 \\ \vdots \\ \Delta\dot{x}_n \end{bmatrix} = \begin{bmatrix} A''_1 & \cdots & 0 \\ \vdots & \ddots & \vdots \\ 0 & \cdots & A''_n \end{bmatrix} \begin{bmatrix} \Delta x_1 \\ \vdots \\ \Delta x_n \end{bmatrix} + \begin{bmatrix} \Delta F_1 \\ \vdots \\ \Delta F_n \end{bmatrix} \begin{bmatrix} \Delta v_1 \\ \vdots \\ \Delta v_n \end{bmatrix} \tag{16}$$

4. Results

In this study, two inverters in parallel are considered for the analysis and have compared the non-linear model presented, the linear model and the electrical simulation model with elementary building blocks. For non-linear modelling, the system of differential equations has been solved for zero initial conditions and steady-state conditions. At a simulation time of 1 second, the system load is increased by 33%. The state vector values before 1 second are used as the initial condition to study the effect of load variation. Similarly, linearised differential equations are also solved for zero initial and steady-state conditions with load perturbation. The linear and non-linear equation system is solved using the RK4 method using the solve_ivp function of the Python script library. Furthermore, the detailed electrical model for the parallel inverter are realized using elementary switching block in Simscape/Simulink MATLAB has been solved using ode23tb solver. The modification with virtual impedance control has been implemented in the two parallel inverters in grid-forming mode performed by Ref. [38] in Simulink. The parameter used in analysis is mentioned in **Table 1**.

The simulation results of active, reactive power, frequency and voltages for both inverters with and without virtual impedances have been mentioned in **Figures 3** and **4**. Similarly, the eigenvalue and frequency response results for the linearised model of the inverter system for load perturbation have been mentioned in **Figure 5**. Additionally, for the frequency and voltage level of individual inverters for different levels of virtual impedance, keeping the base as coupling impedance is mentioned in **Figure 6**.

Parameters	Values	Parameters	Values	Parameters	Values
Voltage (V)	400 V	Voltage Controller Proportional Gain (K_{pv})	0.2858	Inverter 2 voltage droop coefficient $(n2)$	1^{-4}
Load Power	2400 W	Voltage controller Integral Gain K_{iv}	595	Filter Inductance L_f	3.5^{-3}
Load Reactive Power	1800 Var	Current Controller Proportional Gain (K_{pc})	55	Filter resistance R_f	0.1
Pertub Load	800 W, 600 Var	Current Controller Integral Gain (K_{ic})	1570	Filter Capacitor C_f	50 uf
System frequency (f)	50 hz	Inverter 1 frequency droop coefficient (m_1)	5^{-5}	Couping resistances R_{c1}, R_{c2}	0.01, 0.05
Filter cut-off frequency w_f	4π	Inverter 2 frequency droop coefficient (m_2)	2^{-5}	Coupling inductances L_{c1}, L_{c2}	0.35 mH, 0.8 mH
f_{sw}	10 kHz	Inverter 1 voltage droop coefficient (n_1)	3^{-4}	Vdc	800 V

Table 1.
Parameters used in the analysis.

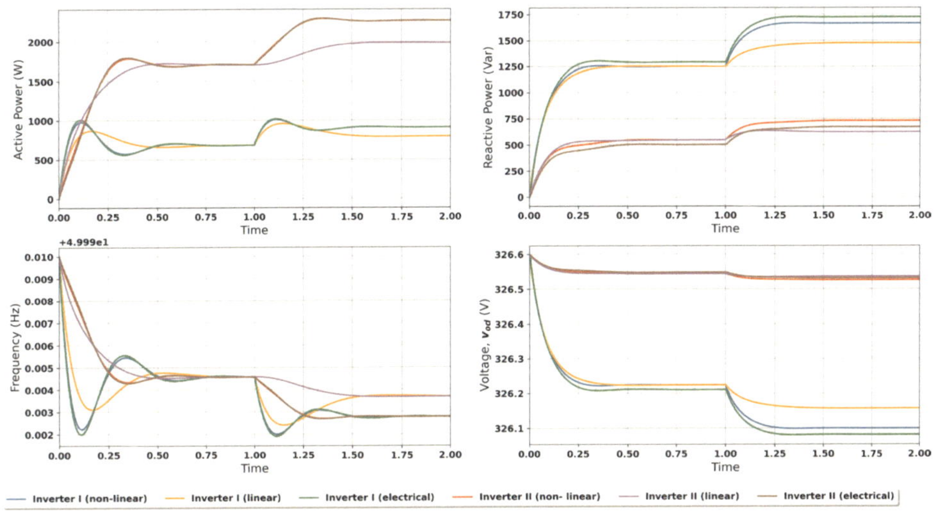

Figure 3.
The inverter power-sharing (top left), reactive power-sharing (top right), frequencies (bottom left) and peak voltages (bottom right) without the virtual impedance control method.

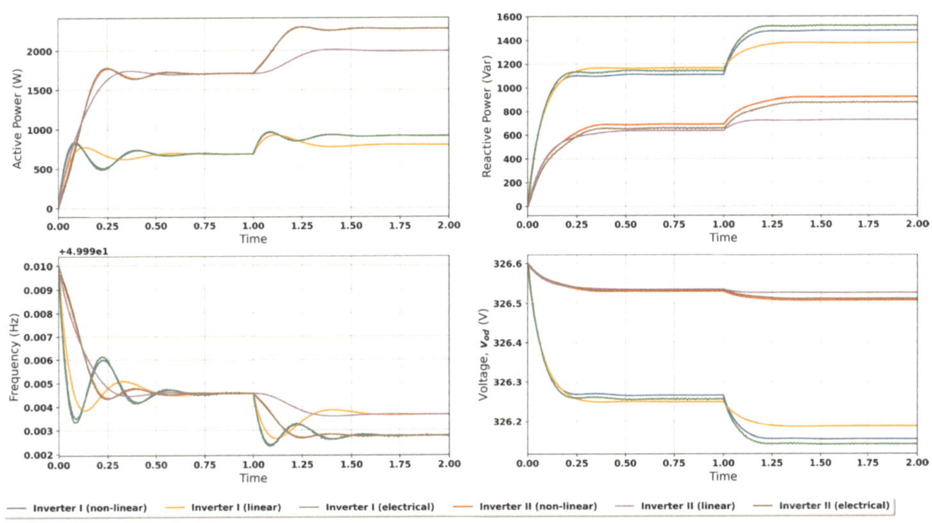

Figure 4.
The inverter power-sharing (top left), reactive power-sharing (top right), frequencies (bottom left) and peak voltages (bottom right) with virtual impedance control method.

A comparison has been shown for the system with and without the virtual impedance is shown in **Figure** 7.

The non-linear and electrical models almost resemble the active power and frequency deviation of individual inverters. However, the electrical simulation shows a slight oscillation. This is due to the actual dynamics of switching devices considered. The linear model results in slight variation from the non-linear and electrical models; however, it represents the overall dynamics of the system. This scenario is identical for both cases, as shown in **Figures 3** and **4**. In the context of power frequency oscillation, using virtual impedance in the system results in an increment of effective

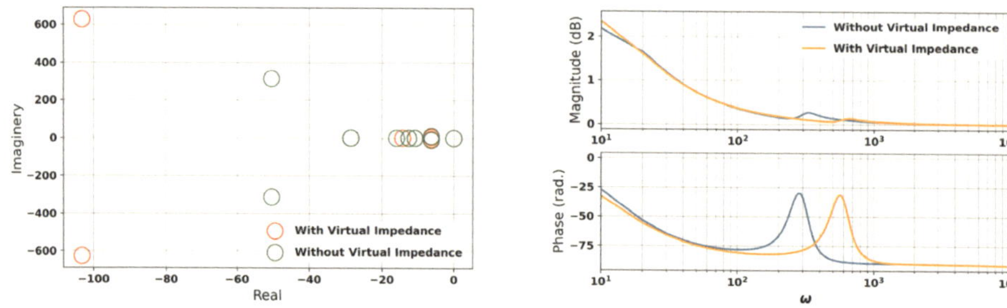

(a) Eigenvalues for steady-state load perturbation

(b) Frequency response for $\frac{\Delta P_1}{\Delta V_{d,ref}}$

Figure 5.
Time domain and frequency domain analysis.

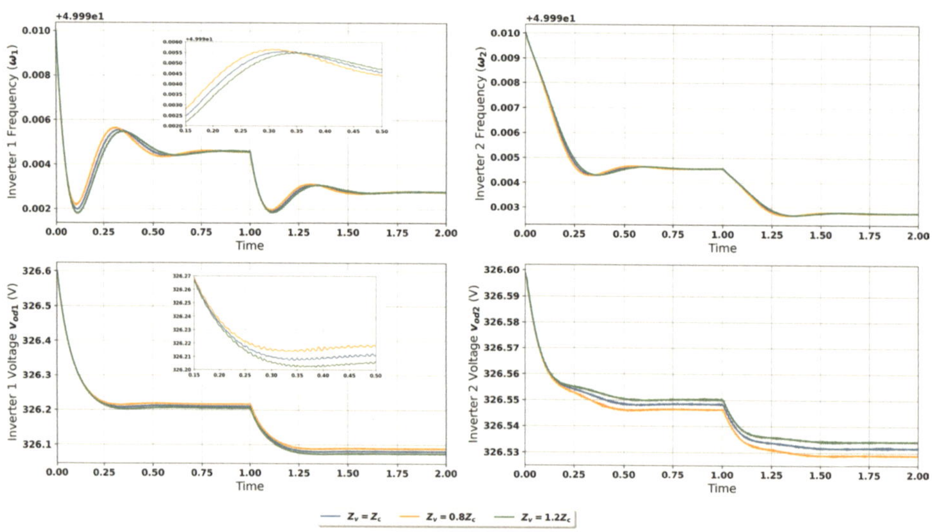

Figure 6.
Voltage and frequency of individual inverters with different levels of virtual impedance performed electrical simulations.

damping in a system with increased overshoot and reduced settling time. The frequency of transient oscillation has been reduced. In the case of voltage and reactive power, the use of a virtual control strategy improves the reactive power sharing and improves the voltage profile of individual inverters. The use of various levels of virtual impedance has been presented in **Figure** 7. Either increment or decrement in the virtual impedance level concerning coupling reactance, the transient condition of the system changes with increasing settling time. Similarly, the voltage level of the system worsens in both cases.

The damping ratio can be calculated using the formula $\zeta = \sigma/\sqrt{\sigma^2 + \omega^2}$, where σ is the real part of an eigenvalue and ω is its imaginary part. By using virtual impedance in the system, the damping ratio of dominant poles increased from 0.1590 to 0.1624

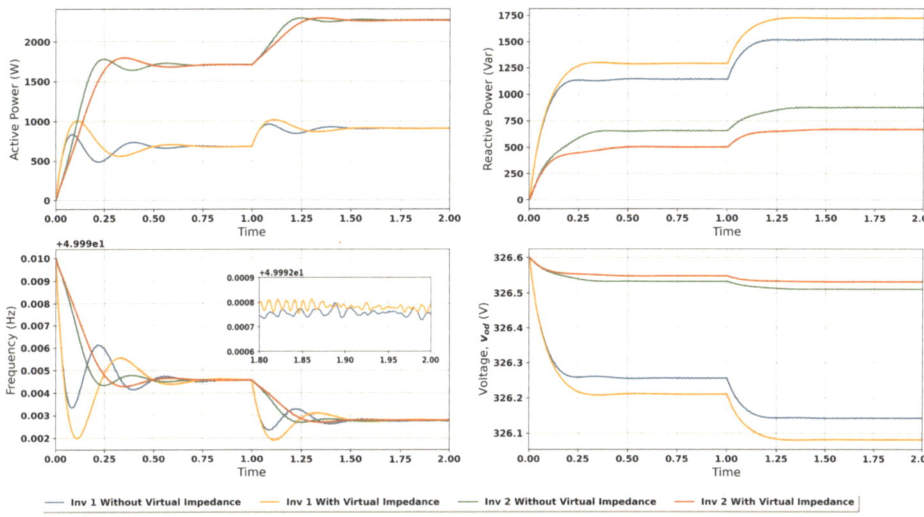

Figure 7.
Electrical simulation of parallel inverter with and without virtual impedances.

and from 0.5194 to 0.6421. Additionally, while there was no change in steady-state active power sharing, reactive power sharing increased from 500 to 600 Var for one inverter and from 1200 to 1250 Var after using virtual impedance. This ability to share higher values of reactive power makes the system more resilient to disturbances and improves voltage stability. However, the presence of higher-order harmonics in the system is still an issue, as shown in **Figure 7**.

Figure 5 a shows the system's eigenvalues for two cases. The location of eigenvalues is shifted slightly to the left-hand side, making the system relatively more stable in the case of the virtual impedance method. Additionally, as suggested by Li et al. [39], when the grid stiffness decreases, that is, weak grid, the resonance frequency shifts to lower frequency bands, suggesting that there could be a chance of low-frequency oscillation. Hence, we can see in a frequency response that the system without virtual impedance possesses a low-frequency resonance frequency compared to the virtual impedance.

5. Conclusion

The study has presented the dynamic model of a parallel connected inverter having virtual impedance control. We have used two voltage source inverters in parallel having droop characteristics. A comparison has been made with and without the virtual impedance control strategy. The non-linear model has been built and is linearised for zero initial condition and steady-state condition to obtain the linearised model using the perturbation technique. The system of the equations has been solved and compared with that of an actual system having elementary switching blocks. Eigenvalue analysis and frequency response analysis were also performed. Results found that using virtual impedance increased the resonant peak of the system and shifted the closed-loop pole to the right-hand side of the s-plane, providing better system stability towards load variation.

Author details

Deependra Neupane* and Nawaraj Poudel
Department of Electrical Engineering, Institute of Engineering, Purwanchal Campus, Tribhuvan University, Dharan, Nepal

*Address all correspondence to: deependra@ioepc.edu.np

IntechOpen

References

[1] Alcala JM, Castilla M, De Vicuña LG, Miret J, Vasquez JC. Virtual impedance loop for droop-controlled single-phase parallel inverters using a second-order general-integrator scheme. IEEE Transactions on Power Electronics. 2010;**25**(12):2993-3002

[2] Simpson-Porco JW, Dörfler F, Bullo F. Synchronization and power sharing for droop-controlled inverters in islanded microgrids. Automatica. 2013; **49**(9):2603-2611

[3] Keyvani-Boroujeni B, Fani B, Shahgholian G, Alhelou HH. Virtual impedance-based droop control scheme to avoid power quality and stability problems in vsi-dominated microgrids. IEEE Access. 2021;**9**:144999-145011

[4] Buraimoh E, Aluko AO, Oni OE, Davidson IE. Decentralized virtual impedance-conventional droop control for power sharing for inverter-based distributed energy resources of a microgrid. Energies. 2022;**15**(12):4439

[5] Kim J-W, Choi H-S, Cho BH. A novel droop method for converter parallel operation. IEEE Transactions on Power Electronics. 2002;**17**(1):25-32

[6] De Brabandere K, Bolsens B, Van den Keybus J, Woyte A, Driesen J, Belmans R. A voltage and frequency droop control method for parallel inverters. IEEE Transactions on Power Electronics. 2007;**22**(4):1107-1115

[7] Tuladhar A, Jin H, Unger T, Mauch K. Control of parallel inverters in distributed ac power systems with consideration of line impedance effect. IEEE Transactions on Industry Applications. 2000;**36**(1):131-138

[8] Borup U, Blaabjerg F, Enjeti PN. Sharing of nonlinear load in parallel-connected three-phase converters. IEEE Transactions on Industry Applications. 2001;**37**(6):1817-1823

[9] Coelho EAA, Cortizo PC, Garcia PFD. Small-signal stability for parallel-connected inverters in stand-alone ac supply systems. IEEE Transactions on Industry Applications. 2002;**38**(2): 533-542

[10] Venkataramanan G, Illindala M. Small signal dynamics of inverter interfaced distributed generation in a chain-microgrid. In: 2007 IEEE Power Engineering Society General Meeting. Tampa, FL, USA: IEEE; 2007. pp. 1-6

[11] Pogaku N, Prodanovic M, Green TC. Modeling, analysis and testing of autonomous operation of an inverter-based microgrid. IEEE Transactions on Power Electronics. 2007;**22**(2): 613-625

[12] Zhang Y, Jiang Z, Yu X. Small-signal modeling and analysis of parallel-connected voltage source inverters. In: 2009 IEEE 6th International Power Electronics and Motion Control Conference. Wuhan, China: IEEE; 2009. pp. 377-383

[13] Zhang X, Liu J, You Z, Zhou L. Small-signal analysis and modeling of parallel inverters based on the droop control method in micro-grid. In: 2013 IEEE Energy Conversion Congress and Exposition. Atlanta, GA, USA: IEEE; 2013. pp. 4580-4586

[14] Parreira WA, Avelar HJ, Vieira JB, Freitas LC, de Freitas LCG, Coelho EAA. Small-signal analysis of parallel connected voltage source inverters using a frequency and voltage droop control including an additional phase shift. Journal of Control, Automation and Electrical Systems. 2014;**25**:597-607

[15] Bennia I, Harrag A, Dailia Y. Small-signal modelling and stability analysis of island mode microgrid paralleled inverters. Journal of Renewable Energies. 2021;**24**(1):105-120

[16] Rasheduzzaman M, Mueller J, Kimball JW. Small-signal modeling of a three-phase isolated inverter with both voltage and frequency droop control. In: 2014 IEEE Applied Power Electronics Conference and Exposition-APEC 2014. Fort Worth, TX, USA: IEEE; 2014. pp. 1008-1015

[17] Rasheduzzaman M, Mueller JA, Kimball JW. An accurate small-signal model of inverter-dominated islanded microgrids using dq reference frame. IEEE Journal of Emerging and Selected Topics in Power Electronics. 2014;**2**(4): 1070-1080

[18] Rasheduzzaman M, Mueller JA, Kimball JW. Reduced-order small-signal model of microgrid systems. IEEE Transactions on Sustainable Energy. 2015;**6**(4):1292-1305

[19] Bryant JS, McGrath B, Meegahapola L, Sokolowski P. Small-signal modeling and stability analysis of paralleled grid-feeding voltage source inverters. In: 2021 IEEE Power and Energy Society General Meeting (PESGM). Washington, DC, USA: IEEE; 2021. pp. 1-5

[20] Mariani V, Vasca F, Guerrero JM. Analysis of droop controlled parallel inverters in islanded microgrids. In: 2014 IEEE International Energy Conference (ENERGYCON). Cavtat, Croatia: IEEE; 2014. pp. 1304-1309

[21] Sharma R, Suhag S. Virtual impedance based phase locked loop for control of parallel inverters connected to islanded microgrid. Computers and Electrical Engineering. 2019;**73**:58-70

[22] Issa W, Al-Naemi F, Konstantopoulos G, Sharkh S, Abusara M. Stability analysis and control of a microgrid against circulating power between parallel inverters. Energy Procedia. 2019;**157**:1061-1070

[23] McGrath B, Mu P, Nazib A, Holmes D, Teixeira C. Small signal dynamic model and stability analysis of a self-synchronizing grid-tied current regulated inverter. In: 2019 IEEE Energy Conversion Congress and Exposition (ECCE). Baltimore, MD, USA: IEEE; 2019. pp. 5561-5568

[24] Unnikrishnan BK, Johnson MS, Cheriyan EP. Small signal stability improvement of a microgrid by the optimised dynamic droop control method. IET Renewable Power Generation. 2020;**14**(5):822-833

[25] Pournazarian B, Seyedalipour SS, Lehtonen M, Taheri S, Pouresmaeil E. Virtual impedances optimization to enhance microgrid small-signal stability and reactive power sharing. IEEE Access. 2020;**8**:139691-139705

[26] Han Y, Qian Z, Shao D, Lin Q, Yin S, Chen M, et al. Small signal stability analysis of microgrid with multiple parallel inverters. In: IOP Conference Series: Earth and Environmental Science. Vol. 687. Zhuhai, China: IOP Publishing; 2021. p. 012112

[27] Guan Y, Vasquez JC, Guerrero JM, Coelho EAA. Small-signal modeling, analysis and testing of parallel three-phase-inverters with a novel autonomous current sharing controller. In: 2015 IEEE Applied Power Electronics Conference and Exposition (APEC). Charlotte, NC, USA: IEEE; 2015. pp. 571-578

[28] Yu K, Ai Q, Wang S, Ni J, Lv T. Analysis and optimization of droop controller for microgrid system based on

small-signal dynamic model. IEEE Transactions on Smart Grid. 2015;7(2): 695-705

[29] Kumar VN, Parida SK. Parameter optimization of universal droop and internal model controller for multi inverter-fed dgs based on accurate small-signal model. IEEE Access. 2019;7: 101928-101940

[30] Leitner S, Yazdanian M, Mehrizi-Sani A, Muetze A. Small-signal stability analysis of an inverter-based microgrid with internal model-based controllers. IEEE Transactions on Smart Grid. 2017; **9**(5):5393-5402

[31] Tan K, So P, Chu Y. Control of parallel inverter-interfaced distributed generation systems in microgrid for islanded operation. In: 2010 IEEE 11th International Conference on Probabilistic Methods Applied to Power Systems. Singapore: IEEE; 2010. pp. 1-5

[32] Guerrero JM, Matas J, de Vicuna LG, Castilla M, Miret J. Decentralized control for parallel operation of distributed generation inverters using resistive output impedance. IEEE Transactions on Industrial Electronics. 2007;**54**(2): 994-1004

[33] Zhong Q-C. Robust droop controller for accurate proportional load sharing among inverters operated in parallel. IEEE Transactions on Industrial Electronics. 2011;**60**(4):1281-1290

[34] Wang S, Su J, Yang X, Du Y, Tu Y, Xu H. A review on the small signal stability of microgrid. In: 2016 IEEE 8th International Power Electronics and Motion Control Conference (IPEMC-ECCE Asia). Hefei: IEEE; 2016. pp. 1793-1798

[35] Wu Y, Wu Y, Guerrero JM, Vasquez JC, Li J. Ac microgrid small-signal modeling: Hierarchical control structure challenges and solutions. IEEE Electrification Magazine. 2019;7(4): 81-88

[36] D'Arco S, Suul JA, Fosso OB. Small-signal modeling and parametric sensitivity of a virtual synchronous machine in islanded operation. International Journal of Electrical Power and Energy Systems. 2015;**72**:3-15

[37] Wang X, Li YW, Blaabjerg F, Loh PC. Virtual-impedance-based control for voltage-source and current-source converters. IEEE Transactions on Power Electronics. 2014;**30**(12): 7019-7037

[38] Sheikhtayyeb MT. Sheikhtayyeb/Centralized-Secondary-Control-in-Inverter-Based-ac-Microgrid. Github; 2024. Available from: https://github.com/sheikhtayyeb/Centralized-secondary-control-in-inverter-based-AC-microgrid

[39] Li C, Yang Y, Cao Y, Wang L, Blaabjerg F. Frequency and voltage stability analysis of grid-forming virtual synchronous generator attached to weak grid. IEEE Journal of Emerging and Selected Topics in Power Electronics. 2020;**10**(3):2662-2671